跟大廚做

北方麵點

Northern
Chinese
Dumplings

李起發 著

萬里機構 · 飲食天地出版社出版

前言

北方人一開口就説："好吃的不如餃子，舒服不如倒着。"意思是最好吃的莫過於餃子，最舒服的事就是躺着。可想而知，餃子在北方人家裏很重要，特別在重要場合都會吃餃子和撈麵款待客人或闔家團聚，有所謂"出門的餃子，回家的麵。"

餃子、包子、小籠包和盒子等，除了好吃外，還蘊含了許多北方飲食文化，夾雜了有趣的風俗和生活智慧。"初一的餃子、初二的麵（拌麵），初三的盒子往家裏轉，初四吃雜燴，初五捏小人嘴（捏餃子）"，因為年初一便開始四處去拜年，閒話家常時少不了張家長、李家短，甚至會説三道四，是非一籠筐。為了封鎖這些是是非非，在年初五上班前便包餃子，象徵緊捏小人嘴，不要到公司胡亂說話。煮水餃時還要特別注意爐火，不能太大火，避免煮破餃子而"露餡了"，讓是非穿出來。

中國人的傳統，要按節令而食。在北方，立春吃春餅，夏至吃拌麵，立秋吃餃子，冬至涮羊肉。立春食春餅是因為春餅有很多餡料，取其餡好和收成好的意思。夏至時於進伏就要吃拌麵，吃拌麵必須拌蒜而食；因俗語有云"不吃蒜不如小米飯"，意思是北方拌麵會拌以很多生吃的配菜，如黃瓜絲、胡蘿蔔絲、香菜碎等等，一旦吃得不對胃，可能要拉肚子，所以要用蒜頭解毒，否則吃了如此多的好東西因不舒服都倒出去，倒不如吃碗小米飯算了。立秋時就要吃以瓜做餡的餃子，飯後吃瓜叫做"咬秋"，暗喻把秋天留下來的意思。冬天到，窗外大雪飄飄，屋裏熱氣騰騰涮羊肉（火鍋），大口大口地吃肉，又大口大口地喝酒，好不美哉！所謂"小寒大寒又一年，月月年年花相似，歲歲年年人不同，有福大家享，有肉大家吃，有美酒大家嚐！"

　本人對麵點製作的心得透過書中食材介紹、製法和份量配方，詳盡解說。只要依從書中指引，讀者們能享受到道地北方口味。再者，書中食譜按照商業配方，公諸同好，絕不藏私。雖然如此，家庭掌廚者，依書製作，一般水餃、鍋貼、蒸包子、蒸饅頭、烙大餅、炸油餅等也難不到大家，與家人一齊做，更增添家庭和諧的歡樂氣氛。

李起發

目錄

包點和花卷

雜類

①工具與常用食材介紹

（A）認識工具才能做好麵點

1 電烙餅器
2 壓麵皮機
3 手絞碎肉器
4 磅
5 中國秤
6 量杯
7 刨絲器
8 各式擀麵杖（棍）
9 油炸用的竹筷子
10 鏟刀

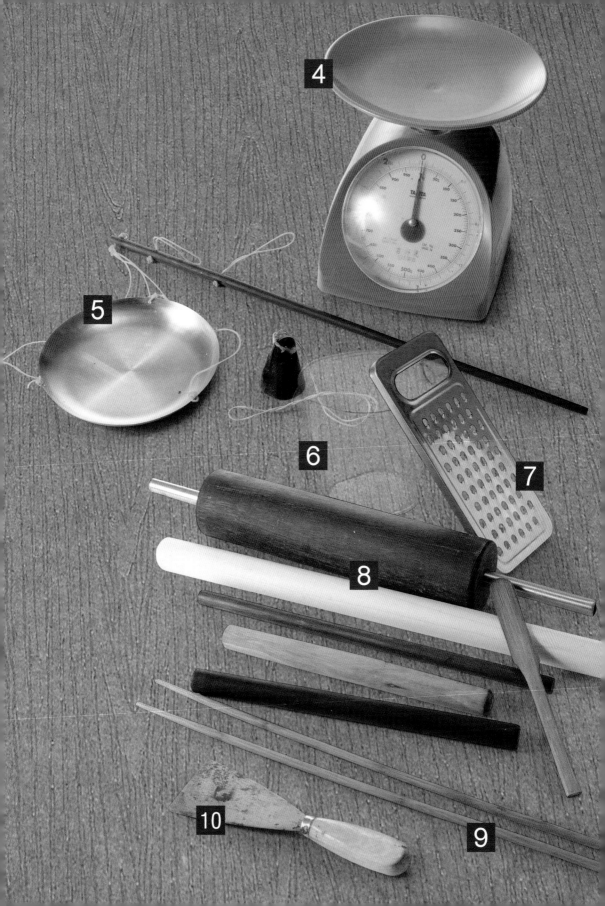

1 不銹鋼蒸籠套煲
2 竹蒸籠
3 煎鍋（平底鑊）

（B）常用食材與刀章運用

（1）常用刀章

好吃的餃子或包子，其餡料的刀章要求纖細或是保有適度嚼口，一般常用的刀章離不開切

1 幼粒、2 幼條、3 中粒、4 粗粒、5 中粗絲、6 指甲片、7 剁碎、8 粗條。

(2) 常用餡的製作

包子、餃子等包餡的製品是否美味，關鍵在於餡料品質和材料的搭配，要確保餡料品質佳，材料搭配合理，就要掌握原材料的性質、特點和最佳炮製方法。

餡料可分三大類

1. 葷餡
   包括豬、牛、羊和雞等肉類，需洗淨、絞碎、剁碎或切丁。

2. 素餡
   包括各類蔬菜、乾菜、蛋類、甜果等，需洗淨、切碎、擠去水份；若是乾菜類就要水發後汆水，再剁碎。

3. 海鮮餡
   包括海產和水產，要去骨、皮和腸臟，洗淨剁碎。若遇到帶有異味的海鮮，如腥、苦、臭等的原料，則要去除異味或用遮蓋的方法為好。

1 牛肉	6 白蘿蔔	11 葱	16 菠菜
2 魚肉	7 南瓜	12 西洋菜	17 鯖魚 / 鮫魚
3 豬肉	8 大白菜	13 京葱	
4 羊肉	9 翠玉瓜	14 韮菜	
5 雞肉	10 小棠菜	15 四季豆	

豬肉餡

■時間：10 分鐘　■份量：1200 克

材料
- 豬上肉600克
- 京葱200克
- 薑20克

調味
- 生油100毫升
- 老抽1湯匙
- 料酒50毫升
- 精鹽1.5茶匙（9克）
- 砂糖2湯匙（24克）
- 味精10克
- 凍水270毫升
- 麻油2湯匙

做法

1 將豬肉洗淨，絞碎，放進盆中，加入料酒、老抽和生油，順方向用力攪動至表面發白。再加入鹽、糖、味料和67.5毫升凍水，攪至黏稠，再將餘下的水份2次加入，攪動至黏稠。

2 加入葱、薑和麻油，攪拌均勻即可。

大廚提示

• 選用豬上肉，瘦肉多而肥肉少，因為有點肥肉的餡料才美味。

• 淨肉餡可以搭配任何切蔬菜，並根據自己口味增減鹹甜味，可做水餃餡。至於包子餡餅的餡就要加入3隻雞蛋和生粉50克（4湯匙），要是添加了100克麵包屑，就可變成獅子頭。

牛肉餡

■時間：10 分鐘　■份量：1200 克

材料
- 鮮嫩牛肉300克，絞碎
- 肥牛片300克，絞碎
- 京葱150克，去皮切碎

調味
- 料酒50毫升
- 精鹽1.5茶匙（9克）
- 砂糖2湯匙（24克）
- 味精10克
- 五香粉1/2茶匙
- 生油150毫升
- 麻油2湯匙
- 凍水150毫升

做法

1 將絞碎牛肉放進盆裏，加進鹽、糖、味料、料酒和生油，順方向用力攪動至黏稠。

2 期間將150毫升清水分3次加入，每次加水時都要打至黏稠，然後再加入葱碎和麻油，拌勻即可。

大廚提示

• 包牛肉冬菇餡餅水餃時，不要忘記加入150克馬蹄肉碎，這樣會令口感更清爽。

• 做牛肉餡最好用前腿肉，再加點肥牛片以添加口感。

羊肉餡

■時間：10 分鐘　■份量：1200 克

材料

- 羊腿肉300克
- 羊腩肉300克
- 京葱200克
- 乾花椒5克

調味

- 生油200毫升
- 精鹽1.5茶匙（9克）
- 砂糖2湯匙（24克）
- 味精10克
- 麻油2湯匙
- 料酒2湯匙
- 凍水200毫升

做法

1 分別將羊腿和羊腩洗淨、絞碎；將京葱去皮、洗淨剁碎。

2 將肉碎、鹽、糖、味料、料酒和生油放於盆裏，倒入過篩的花椒粉，用力順一方向攪至黏稠。

3 加入1/3水，再攪至黏稠，再分2次加入餘下水份，最後加葱碎和麻酒，拌勻即可。

大廚提示

- 羊肉多為紅肉，羶味很重，宜用花椒去羶。

- 建議用15克花椒粒加400毫升清水，上火煲10分鐘，放涼，篩去花椒粒，取代材料中的200毫升凍水做湯底，配上西葫蘆（翠玉瓜）、津白和白蘿蔔，便成津白粉絲羊肉丸子湯。

- 這餡不宜搭配韭菜，味道不搭。

雞肉餡

■時間：10 分鐘　■份量：1000 克

材料

- 雞腿肉300克
- 雞胸肉300克
- 京葱白100克
- 薑20克，切碎

調味

- 生油150毫升
- 料酒50毫升
- 精鹽1茶匙（6克）
- 砂糖2湯匙（24克）
- 雞粉1湯匙（15克）
- 凍水100毫升
- 麻油2湯匙
- 蛋白2隻

做法

1 將雞肉洗淨、絞碎或剁碎；將京葱去皮、洗淨剁碎。

2 將肉碎、鹽、糖、雞粉、料酒、生油和蛋白放於盆裏，加50毫升水，順方向攪至黏稠。

3 再加50毫升水打至黏稠，拌入葱白碎和麻油即可。

大廚提示

- 最好選用胸脯肉加腿肉，白肉沒肥油而紅肉就有肉味，混合使用效果更好。

- 這餡的質地細膩，老人家食用最好，宜包餛飩。若做丸子湯，就加2茶匙三花奶混合，味道和色澤都會更好。

- 雞肉餡不易搭配太過粗糙的蔬菜，以薺菜、芽菜、豆苗、翠玉瓜和老黃瓜為佳。

魚肉餡

材料

- 鮫魚1條（約2斤，1200克）
- 鮮薑50克
- 清水100毫升
- 冰粒100克
- 蛋白6隻
- 胡椒粉2茶匙

調味

- 生粉4湯匙（50克）
- 魚露1湯匙
- 鹽1茶匙（6克）
- 砂糖1湯匙（12克）
- 白醋1湯匙（15毫升）
- 生油100毫升
- 豬晶油*100克

* 豬晶油：晶牌豬油，優質精煉豬油，產自荷蘭。

做法

1 把薑、水和冰粒一起放進攪拌機中，打成薑汁。

2 將鮫魚洗淨、去頭、去骨、去皮，餘下約550-700克肉左右。

3 為了防止纖細魚骨傷喉，用絞碎機把魚肉反覆絞2次，才放進所有材料和調味，順方向攪打黏稠即可。

大廚提示

- 經驗豐富者，能取出較多魚肉。
- 與鮫魚肉最搭配是韭菜或鹹酸菜。
- 魚肉製餡一定要選擇魚骨少的新鮮魚，如比目魚、桂花魚、生魚、鱸魚和鮫魚。

素食餡

材料

- 老黃瓜2條（約3.5斤，2100克左右）
- 雞蛋5隻，炒熟，切碎
- 水發木耳100克，切碎
- 京葱100克，洗淨切碎
- 生油150毫升
- 精鹽1茶匙（6克）
- 砂糖2湯匙（24克）
- 味精10克
- 麻油2湯匙

做法

1 首先將老黃瓜去皮、去瓤、洗淨，切成絲後撒入2茶匙鹽，拌勻，醃5分鐘。

2 擠去老黃瓜水份，剩約500-600克左右瓜肉。

3 把老黃瓜放在砧板上切幾刀，放進盆裏，加進所有材料，拌勻即可。

大廚提示

- 素餡，即無葷，它可分大無葷與小無葷。前者的規範比較講究，包括海產、薑、葱、蒜、韭菜、韭黃和蛋類都不能用；小無葷則只是沒有肉而已。
- 素餡材料很講究，如乾貨、蘑菇、木耳、黃花菜（金針菜）、髮菜和乾菜芯（菜乾）等等，經發水後使用；有些乾菜還要用水煮入味後才可使用。

② 不可不學的麵糰，一定要學識

世上沒有學不好的廚藝。北方麵點易學難精，要學做北方菜，第一步就要先學會做麵糰，而要做好麵糰，最主要是看能否掌握到各類麵粉的性能與用途，讓你按扁搓圓，隨心所欲。學會做麵糰，便有了根基穩固的基本理念，再重複試煉，融會貫通，功夫自然到家。

麵粉分為三種，高筋麵粉含 12%-15% 蛋白質，中筋麵粉含 9%-11% 蛋白質，低筋麵粉則含 6%-9% 左右的蛋白質，脂肪含量越高，遇水後筋性就越大，所以高筋麵粉不宜用於家常菜。至於中筋麵粉和低筋麵粉，則要根據自己的能力和需求去選擇。

發酵麵糰

■時間：1小時　■份量：1000克

大廚提示

- 在麵粉中加入一定比例的酵母菌與清水的合成麵糰，就是發酵麵糰。
- 酵母菌能使麵糰膨脹，在發酵過程中會產生酒和酸的氣味，這時的麵糰含酸性，故在搓揉麵糰時，需加入少許糖或食用鹼，以中和酸味和酸鹼度。
- 發酵速度的快或慢，取決於酵母菌的用量和室溫的高低。
- 發酵麵糰有時會用上老麵（麵種或麵肥），至於它的用量多少，發酵時間的長短，室溫的高低均直接影響到發酵效果。含麵種的麵糰，其發酵過程比較長，氣味濃烈，需要用食用鹼以中和酸鹼度，用量就全憑個人經驗決定，但對經驗不足的人士，難於控制。
- 麵糰最少要在用前的2小時開始做。
- 若室溫低於攝氏20度的話，醒麵時間就要延長20分鐘。

材料

- 麵粉600克
- 依士（酵母）1茶匙
- 泡打粉（發粉）2茶匙
- 砂糖2湯匙
- 生油2湯匙
- 清水300毫升

做法

1 將麵粉、依士、泡打粉、砂糖和生油拌勻。

2 再加清水300毫升分3次加入，一邊加入一邊拌勻，搓至表面光滑。

3 用保鮮紙密蓋麵糰，置室溫待50分鐘，再搓揉10分鐘，便成生饅頭坯，待發酵（醒發）20分鐘。

4 將麵糰放上蒸鍋，用大火蒸20分鐘，即成熟饅頭。

a

b

c

d

半發酵麵糰

■時間：10分鐘　■份量：1000克

材料

- 麵粉600克
- 依士1茶匙
- 砂糖1湯匙
- 生油1湯匙
- 雞蛋1隻
- 常溫水350毫升

做法

1. 在麵粉裏加入依士、砂糖、生油和雞蛋混和，拌勻。

2. 加入常溫水，邊加水，邊攪拌至成絮狀，搓揉成表面光滑的麵糰，用濕布蓋好，醒發10分鐘即可。

3. 下劑子（撕成小麵糰，本地叫"出體"），按扁擀圓成皮，便可應用了。

大廚提示

- 這種麵糰適合做餡餅。

- 因為煎餡餅是不加水的，所以調製麵糰時宜軟不宜硬。

- 煎時也不要太大火，以免外皮已煎焦，但餡料卻未熟。

冷水麵糰

材料

- 麵粉600克
- 清水300毫升

做法

1 將清水300毫升分3次加入麵粉中，拌勻。

2 搓揉至表面光滑，用保鮮紙密蓋（蓋嚴），置室溫待（北方行內術語稱"醒"，港式行內術語稱為"恆身"）20分鐘即可。

大廚提示
- 建議非專業人士在麵盆裏和麵。若麵糰用來做水餃，要硬一點；如果是做烙餅，便要軟一點。

a

b

c

燙麵糰

材料

- 麵粉600克
- 生油2湯匙
- 滾水500毫升

做法

1 把生油加到麵粉裏，快速倒入500毫升滾水，用筷子將盆中的乾粉拌勻。

2 待差不多成糰時，再用手揉搓，小心燙手，便成燙麵糰，然後用刀在其上面劃兩、三刀以便散熱。

大廚提示
- 根據用途可以改變麵粉與滾水的比例。
- 燙麵製品，一定要配合冷水麵糰使用，目的是調節燙麵糰的軟硬度，至於冷水麵糰用量的比例（配比）會直接影響到製品的效果。

a

b

c

d

香菇牛肉
水餃

材料

冷水麵糰
- 低筋麵粉600克
- 清水280毫升
- 精鹽1/4茶匙

餡料
- 牛肉餡400克（請參閱第12頁）
- 洋葱1個，去衣切碎
- 發水（即已浸泡）冬菇200克，切粒
- 麻油2湯匙，後下

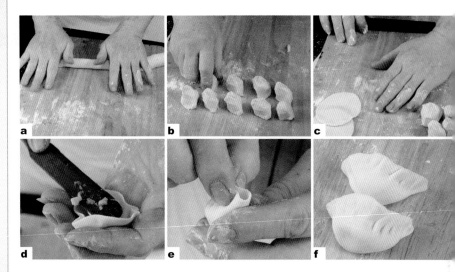

做法

1 把所有餡材料混合，再加入麻油拌均勻，備用。

2 冷水麵糰的做法，請參閱第17頁，醒發15分鐘。

3 取出部份麵糰，揉搓成如拇指般粗細的條狀，下成劑子（即分體，撕下的小粉糰粒），每個5-6錢（19-22克）重，按扁擀長，最好週邊薄而中間厚。

4 將麵皮平放，用餡挑將適量餡料放於麵皮上，壓平。

5 把麵皮對摺，貼合捏緊，順序推摺或捏緊，便成水餃生坯。

6 燒滾水，用鍋勺推動滾水至起漩渦，放入生坯後輕輕攪動，不讓水餃黏底。

7 蓋鍋蓋，待水滾時揭開鍋蓋，倒入一碗清水，再蓋鍋蓋，水再翻滾後時揭蓋，用中火續煮2分鐘即可。

大廚提示

- 煮水餃另一方法，就是水滾放入生坯，用勺輕輕順方向推至它在水中轉動，蓋上鍋蓋，待水再滾，不要再蓋上鍋蓋，續煮7分鐘左右，撈出一隻，以手指捏一下，餃皮不黏手、肉餡變硬時，沖入1碗冷水，再滾起即可。

- 素餡的烹煮時間很短，不能用手捏測試水餃熟度，眼觀水餃麵皮脹起，就改用小火煮1分鐘左右就熟了。

- 北方人家每年在臘八時做臘八醋，材料有陳醋3斤（1800克），大蒜瓣1斤（600克），置陰涼地方，靜置一旁浸泡至農曆年三十，並在除夕夜吃餃子時作蘸汁。

韭菜豬肉
水餃

大廚提示

• 韭菜性偏寒，食用時不宜飲用冰凍飲品。

• 這種蔬菜有清理腸道作用，故腸胃不好者宜配白大米粥享用。北方人稱大米為白米。

材料

冷水麵糰

- 麵粉500克
- 清水250毫升

餡料

- 豬肉餡400克（請參閱第12頁）
- 韭菜500克，切碎

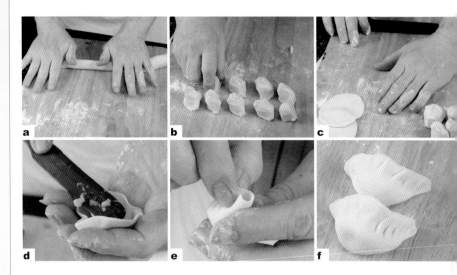

做法

1 冷水麵糰的做法，請參閱第17頁。

2 餡料拌勻備用。

3 取出部份麵糰，揉搓成如拇指般粗細的條狀，下成劑子，每個5-6錢（19-22克）重，按扁擀長，最好週邊薄而中間厚。

4 將麵皮平放，用餡挑將適量餡料放於麵皮上，壓平。

5 把麵皮對摺，貼合捏緊，順序推摺或捏緊，便成水餃生坯。

6 燒滾水，用鍋勺推動滾水至起漩渦，放入生坯後輕輕攪動，不讓水餃黏底。

7 蓋鍋蓋，待水滾時揭開鍋蓋，倒入一碗清水，再蓋鍋蓋，水再翻滾後時揭蓋，用中火續煮2分鐘即可。

羊肉水餃

材料

冷水麵糰	餡料
● 麵粉500克	● 羊肉餡600克（請參照第13頁）
● 清水250毫升	● 麻油2湯匙，後下

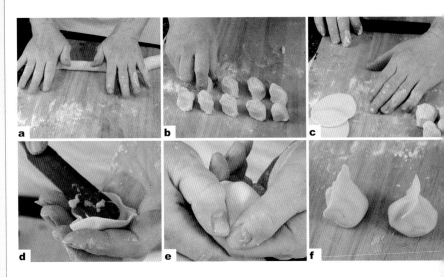

做法

1 把所有餡材料混合，再加入麻油拌均勻，備用。

2 冷水麵糰的做法，請參閱第17頁，醒15分鐘。

3 取出部份麵糰，揉搓成如拇指般粗細的條狀，下成劑子，每個5-6錢（19-22克）重，按扁擀長，最好週邊薄而中間厚。

4 將麵皮平放，用餡挑將適量餡料放於麵皮上，壓平。

5 把麵皮對摺，貼合捏緊，順序推摺或捏緊，便成水餃生坯。燒滾水，用鍋勺推動滾水至起漩渦，放入生坯後輕輕攪動，不讓水餃黏底。

6 蓋鍋蓋，待水滾時揭開鍋蓋，倒入一碗清水，再蓋鍋蓋，水再翻滾後時揭蓋，用中火續煮2分鐘即可。

大廚提示

● 羊肉羶味重，很多人吃不慣，為了保持獨特風味，可用花椒改變羶味。北方人會把花椒炒焦磨粉，拌入羊肉拌醃減羶味。另把花椒粒加水煮出香麻的花椒水，與羊肉混合，十分匹配。

三鮮盒子

材料

冷水麵糰

- 麵粉500克
- 清水250毫升

餡料

- 豬上肉300克，絞碎
- 鮮蝦仁200克，瀝乾
- 水發木耳50克，切碎和後下
- 韭菜200克，切碎
- 油50毫升，拌韭菜用
- 雞蛋2隻，炒熟切碎和後下

餡調料

- 生抽3湯匙
- 砂糖1湯匙
- 鮑魚汁1湯匙
- 花生油150毫升
- 冷水100毫升
- 麻油2湯匙

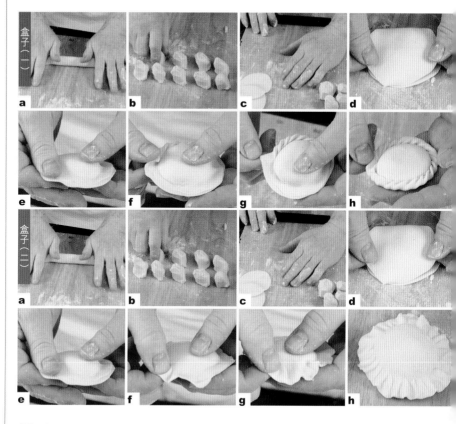

做法

1. 冷水麵糰的做法，請參閱第17頁，靜待15分鐘。

2. 絞碎豬肉加入生抽、糖、花生油、鮑魚汁和50毫升冷水，攪打至黏稠，再加入剩餘的50毫升冷水，續攪至黏稠，加入韭菜碎、雞蛋碎和麻油，拌勻，再加入其他材料，拌勻，置放入雪櫃備用。

3. 取出部份麵糰，揉搓成如拇指般粗細的條狀，下成劑子，每個5-6錢（19-22克）重，按扁擀長，最好週邊薄而中間厚。

4. 將麵皮平放，用餡挑將適量的餡放於麵皮上，再鋪另一片麵皮，貼合捏緊，沿週邊扭紋鎖邊，狀如盒子，便成水餃生坯。

5. 燒滾水，用鍋勺推動滾水至起漩渦，放入生坯後輕輕攪動，不讓水餃黏底。

6. 蓋鍋蓋，待水滾時揭開鍋蓋，倒入一碗清水，再蓋鍋蓋，水再翻滾後時揭蓋，用中火續煮2分鐘即可。

津白三鮮
水餃

材料

冷水麵糰

- 麵粉500克
- 清水250毫升

餡料

- 豬肉餡300克，
  （請參照第12頁）
- 鮮蝦仁300克，瀝乾
- 津白碎200克
- 炒雞蛋碎2隻
- 水發木耳碎100克

调味

- 精鹽1/4茶匙
- 麻油1湯匙
- 鮑魚汁1湯匙

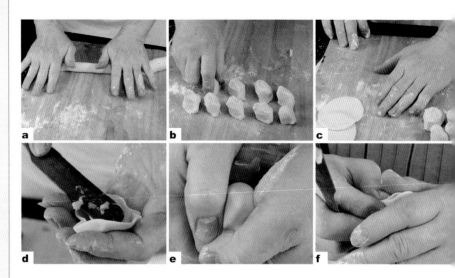

做法

1 冷水麵糰的做法，請參閱第17頁，靜待10-20分鐘。

2 將所有餡料混合，加入調味攪勻，放入雪櫃備用。

3 取出部份麵糰，揉搓成如拇指般粗細的條狀，下成劑子，每個5-6錢（19-22克）重，按扁擀長，最好週邊薄而中間厚。

4 將麵皮平放，用餡挑將適量餡料放於麵皮上，壓平。

5 把麵皮對摺，貼合捏緊，順序推摺或捏緊，便成水餃生坯。

6 燒滾水，用鍋勺推動滾水至起漩渦，放入生坯後輕輕攪動，不讓水餃黏底。

7 蓋鍋蓋，待水滾時揭開鍋蓋，倒入一碗清水，再蓋鍋蓋，水再翻滾後時揭蓋，用中火續煮2分鐘即可。

大廚提示

- 這款餃子不能以擠壓法測試檢驗熟的程度，需觀察烹煮者過程的水餃內產生的蒸汽膨脹起來，再加入兩碗清水。
- 食用時，最好砸碎大蒜變成蒜泥，加2湯匙醋拌勻，再混少許麻油作蘸醬，甚具特色。

韭菜魚肉水餃

Chives and Fish Dumplings

材料

冷水麵糰

- 麵粉500克
- 清水250毫升

餡料

- 魚肉餡300克（請參閱第14頁）
- 韭菜300克，切碎
- 薑蔥油2湯匙

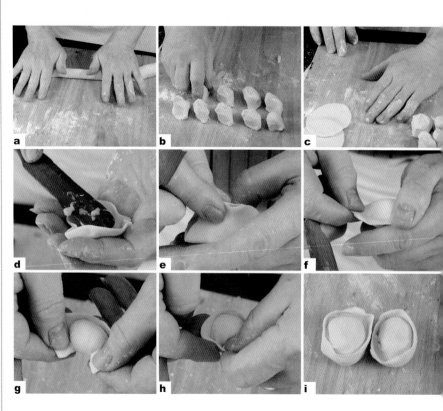

做法

1 冷水麵糰的做法，請參閱第17頁，靜待10-20分鐘。

2 把所有材料一起攪拌均勻，放雪櫃備用。

3 取出部份麵糰，揉搓成如拇指般粗細的條狀，下成劑子，每個
  5-6錢（19-22克）重，按扁擀長，最好週邊薄而中間厚。

4 將麵皮平放，用餡挑將適量的餡放於麵皮上，對摺壓平，並將兩
  端貼合一起，造成環形，便成水餃生坯。

5 燒滾水，用鍋勺推動滾水至起漩渦，放入生坯後輕輕攪動，不讓
  水餃黏底。

6 蓋鍋蓋，待水滾時揭開鍋蓋，倒入一碗清水，再蓋鍋蓋，水再翻
  滾後時揭蓋，用中火續煮2分鐘即可。

大廚提示

- 這水餃在烹煮時會膨脹，看見此情況就轉用小火。
- 不吃重口味的韭菜，可改為魚肉芹菜餡，材料是魚肉和中國芹菜各300克，但芹菜不用焯，確保鮮美風味。
- 改用香菜（即芫荽），只要洗淨切碎拌勻，食味也很不錯。

花素鍋貼

材料

合成麵糰

- 燙麵糰300克（請參閱第17頁）
- 冷水麵糰300克（請參閱第17頁）

餡料

- 小棠菜／上海白菜1200克
- 水發木耳100克，切碎
- 京葱100克，切碎
- 雞蛋4隻，炒好切碎

餡調味

- 精鹽1茶匙
- 豬晶油50克
- 生油80毫升
- 雞粉1茶匙
- 鮑魚汁1湯匙
- 麻油1湯匙，後下

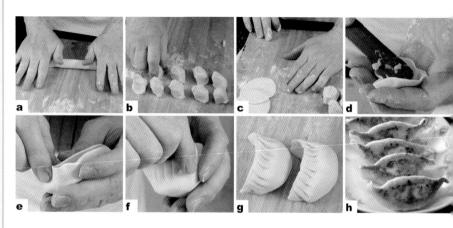

做法

1 將所有餡料混合，加入調味拌勻，最後淋上麻油拌勻，放雪櫃備用。

2 將燙麵糰和冷水麵糰揉合均勻，一邊揉一邊加入適量乾麵粉，好使麵糰表面光滑。

3 取出部份麵糰，揉搓成如拇指般粗細的條狀，下成劑子，每個5-6錢（19-22克）重，按扁擀長，最好週邊薄而中間厚。

4 將麵皮平放，用餡挑將適量餡料放於麵皮上，壓平。

5 把麵皮對摺，貼合捏緊，順序推摺或捏緊，便成水餃生坯。將鍋貼生坯按同一個方向整齊擺放在煎鍋裏，用大火煎1分鐘，再加入300毫升清水，蓋上鍋蓋。

6 燒至水幾近乾時，加入2湯匙生油，蓋上鍋蓋，轉中火，不斷轉動煎鍋，使鍋貼均勻受熱，待水份被蒸發後，鍋貼煎至金黃，可鏟起上碟。

大廚提示

- 當包鍋貼的技巧熟練後，可將手法改為抱拳式，即是用左手的拇指、食指和右手的拇指、食指，一齊擠壓的包製方法。
- 如果用擠壓法包鍋貼，會擠出蔬菜的水份，所以麵皮要多加水份，燙麵糰的份量要減少，否則鍋貼會出現皮大餡小的現象。

羊肉翠瓜
鍋貼

材料

合成麵糰

- 燙麵糰300克（請參閱第17頁）
- 冷水麵糰400克（請參閱第17頁）

餡料

- 羊肉餡400克（請參閱第13頁）
- 翠玉瓜1.5斤（900克），刨絲
- 精鹽1茶匙

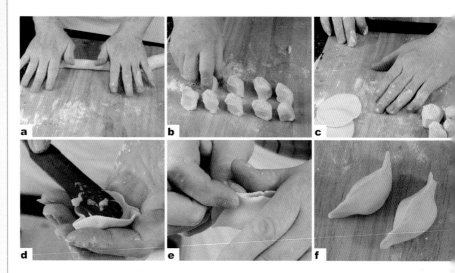

做法

1　將翠玉瓜絲拌入精鹽，攪勻，醃3分鐘後擠去水份，用刀略切幾下，放入羊肉餡裏，拌勻，放雪櫃備用。

2　把麵糰揉搓成如拇指般粗細的條狀，下成劑子，每個5-6錢（19-22克）重，按扁擀圓成皮，放進約1湯匙餡料。

3　把鍋貼皮覆上，推摺花紋兼貼合一起，便成鍋貼生坯。

4　燒熱鍋，放油，按同一個方向整齊排放鍋貼，加入300毫升清水，蓋上鍋蓋，用大火煮1分鐘。

5　燒至水份接近收乾時，加入2湯匙生油，蓋上鍋蓋，轉中火，不斷轉動煎鍋，使鍋貼受熱均勻，煎至金黃色，可上碟。

大廚提示

- 翠玉瓜是長江以北的一種西葫瓜，外形與香港的翠玉瓜相似，任其生長，大如冬瓜。現因經濟效益，北方大酒樓偶有羊肉西葫鍋貼，但小酒樓就改用翠玉瓜了。

傳統鍋貼

材料

合成麵糰
- 燙麵糰100克
  （請參閱第17頁）
- 冷水麵糰500克
  （參閱第17頁）

餡料
- 牛肉餡600克
  （請參閱第12頁）
- 洋葱碎200克

水晶皮料
- 清水300毫升
- 麵粉1/2湯匙
- 鹽1/3茶匙

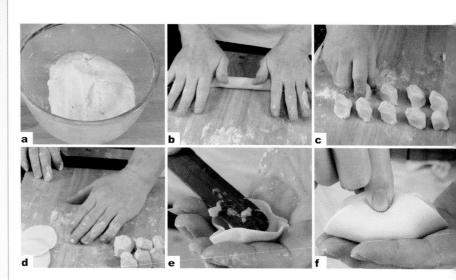

做法

1. 將牛肉餡和洋葱拌勻，放雪櫃凝固，待用。

2. 將燙麵糰和冷水麵糰揉合均勻，一邊揉一邊加入麵粉，好使麵糰表面光滑。

3. 從麵糰中取出25克，按扁後擀圓成皮，放進20克餡料。

4. 覆上麵皮，對摺後兩端向內推入，捏實頂位，便成鍋貼生坯。

5. 水晶皮糊材料拌勻，備用。

6. 將鍋貼排進鍋裏，用大火煎1分鐘左右，加進水晶皮料的300毫升水，蓋上鍋蓋，燒至水份接近收乾，轉中火不斷轉動煎鍋，使鍋貼受熱均勻，1分鐘後沿鍋邊倒下麻油1湯匙。

7. 轉動鍋身，使所有鍋貼一起活動，再用盤子扣在煎鍋上，煎至鍋貼皮成水晶皮，即按住碟子反過煎鍋便可。

大廚提示
- 為免燙傷，沒有操作經驗人士不要這樣以碟子作蓋，可以按一般做法，蓋上鍋蓋直至煮好即可。
- 傳統的鍋貼最好是全肉餡或肉多菜少，由於不封口，全菜餡會漏出來。

天津
水麵餅

北方
豆腐腦

材料

- 原味熱豆腐花350克
- 肉碎50克
- 鮮蝦仁4粒
- 水發木耳3朵
- 水發黃花菜6-8條
- 麵筋1塊，切5-6塊
- 雞蛋1隻，打散
- 油條1/4條，切小段

調味

- 上湯300毫升
- 老抽，調色用
- 精鹽1/3茶匙
- 味精1/3茶匙
- 紅椒油1茶匙
- 水調芝麻醬1湯匙
- 蒜泥水1/2茶匙
- 生粉水2湯匙

做法

1. 起鍋熱油，炒香肉碎和蝦仁，加入上湯和老抽調成醬紅色。
2. 加入精鹽、味精、麵筋、木耳、黃花菜，以大火煮香，放入生粉水，再淋雞蛋液，撒上油條塊即可。
3. 用個大碗，先放熱豆腐花，再用勺打上滷料，後淋上水調芝麻醬、紅椒油、蒜泥水。

材料

- 麵粉600克
- 清水350毫升
- 花生油1湯匙，夾層掃油用

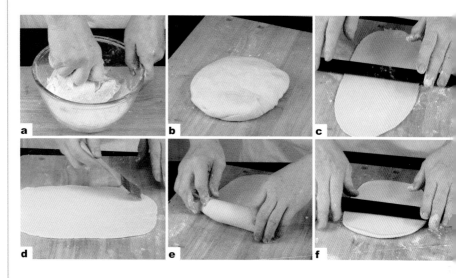

做法

1. 麵粉開穴，一邊加入清水；一邊攪拌成絮狀，搓揉成表面光滑麵糰，然後用保鮮紙或用濕布蓋好，醒發20分鐘，再次搓滑麵糰至光滑，蓋好，醒發10分鐘，備用。

2. 在麵板上撒上乾麵粉，放上已醒發的麵糰，用擀麵杖擀成薄片。

3. 掃上花生油，由外向內捲起，根據所需大小切成幾段。

4. 將麵坯的兩端封好，按扁，擀至1公分厚的圓餅生坯。

5. 燒熱鍋，淋下油，放入餅坯，煎至兩面金黃即可。

大廚提示

- 在烙餅的過程中，用鍋鏟輕壓麵餅的邊緣，讓餅內產生蒸汽，繼而令氣體受熱膨脹變成球狀，成餅後內裏層次分明。

- 可用津白絲和豬肉絲與餅同炒，成炒餅；或是餅切絲，放於燒滾湯中，即成燴餅或湯餅。

津白肉絲
燜餅

材料

餅

- 麵粉600克
- 清水350毫升
- 花生油1湯匙，夾層掃油用

配料

- 豬肉絲300克
- 天津白菜半棵
- 水發冬菇6朵
- 京葱碎100克
- 上湯400毫升

調味

- 鹽適量
- 味精適量
- 老抽適量

 a 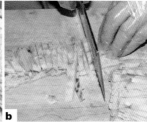 b  c

做法

煎家常餅

1 麵粉開穴，一邊加入清水；一邊攪拌成絮狀，搓揉成表面光滑麵糰，然後用保鮮紙或用濕布蓋好，醒發20分鐘，再次搓滑麵糰至光滑，蓋好，醒發10分鐘，備用。

2 在麵板上撒上乾麵粉，放上已醒發的麵糰，用擀麵杖擀成薄片。

3 掃上花生油，由外向內捲起，根據所需大小切成幾段。

4 將麵坯的兩端封好，按扁，擀至1公分厚的圓餅生坯。

5 燒熱鍋，淋下油，放入餅坯，煎至兩面金黃即可。

燜餅

1 先將家常餅1開3，再切成絲條，備用。

2 洗淨天津白菜，橫切粗絲；水發冬菇切絲，京葱切碎。

3 燒熱鍋，略炒葱碎，放下肉絲炒香，再放進冬菇絲和津白絲，用大火炒軟。

4 加入調味燒至大滾，放入餅絲快速炒勻收汁即可。

大廚提示

- 這款簡單燜餅，在七十至八十年代，北方大小飯店食肆的菜牌必備，更是打工一族的人氣主食。

葱花餅

材料

麵糰

- 麵粉600克
- 生油1湯匙
- 清水350毫升

配料

- 豬網油或肥肉片適量
- 葱花2-3湯匙
- 鹽1/2茶匙
- 五香粉1/4茶匙

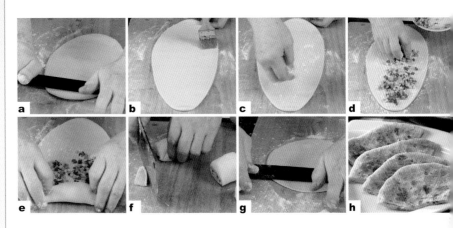

做法

1 將麵粉加入生油調勻，再徐徐加入清水攪拌成絮狀，搓揉成表面光滑的麵糰，蓋上保鮮紙或用濕布蓋好，醒發20分鐘，再揉至表面光滑，蓋上保鮮紙或濕布，待10分鐘。

2 在麵板上撒上乾麵粉，放上已醒好的麵糰，用擀麵杖擀成薄片。

3 掃上花生油，放上豬網油或肥肉片，再撒上葱花、鹽和五香粉，由外向內捲起，按所需大小，切數段。

4 將單段麵坯的兩端封好，按扁，擀至1公分厚的圓餅生坯。

5 用文火燒熱餅鍋，倒入50毫升花生油，放入圓餅生坯，煎烙至兩面金黃，取出，瀝油即可。

大廚提示

- 在北方，一般會將葱花家常餅配鴨湯小餛飩。所謂小餛飩即餡少皮大，好湯頭。材料：免治瘦豬肉300克，鹽2茶匙，味精1茶匙，砂糖1湯匙，麻油20毫升，所有材料拌勻，包上薄薄餛飩皮，可做100粒。

單餅或
北京填鴨餅

北京片皮鴨

材料

- 北京填鴨1隻

上皮料

- 麥芽糖2湯匙
- 白醋200毫升
- 溫水300毫升
- 天津玫瑰露酒1湯匙

做法

1. 填鴨洗淨，在右翅膀下切開1吋左右，從刀口取出鴨內臟及喉管，洗淨，以左手封住刀口，並由鴨的肛門充氣，使鴨皮脹起。

2. 用吊勾掛起全鴨，燒滾水，並將勾好的填鴨放在滾水中焯3-5秒鐘，出水。

3. 用溫水溶化麥芽糖和白醋，淋在填鴨全身，掛在通風處，用電風扇吹乾，使鴨皮收緊。

4. 放入烤爐，以中火（220℃）烤45分鐘左右即可。

材料

合成麵糰

- 燙麵糰400克（請參閱第17頁）
- 冷水麵糰100克（請參閱第17頁）
- 麵粉適量，作補（即粉培）用

a b c
d e f

做法

1. 將冷水麵糰和燙麵糰之間撒上麵粉，混合均勻，搓至表面光滑，蓋上保鮮紙或用濕布蓋好，醒發15分鐘，便成單餅麵糰。

2. 在麵板或工作枱撒上麵粉，放上麵糰，搓成條，再分成小糰，每糰20克，擀成單餅生坯。

3. 用中火燒熱餅鍋，起首時可放少許油，然後放進單餅生坯，烙至兩面變色和起泡即可。

大廚提示

- 烙餅不要用豬油，宜用生油，否則單餅遇冷後容易變硬。

- 其特點是薄如紙，軟如綿，放在報紙上可見到標題字。

- 如果在家中製作北京填鴨，需要借助烤箱烤熟，就要燒熱油淋鴨身上色了。

葱油餅

材料

麵糰

- 麵粉600克
- 生油2湯匙
- 清水360毫升

餡料

- 香葱肉600克，切碎
- 豬晶油200克，麻油25克
- 精鹽1茶匙，味精1湯匙

大廚提示

- 包葱油餅時要盡量包緊，以免葱汁外溢，影響外觀。
- 全油酥麵的葱油餅是美國玫瑰牌麵粉10斤（6000克）、金像根麵粉2斤（1200克）、用豬晶油5斤（3000克）和雞蛋7隻混合，搓揉成光滑的麵糰；但注意一定先加入水份才可開動機器，反之則不達。

做法

1　麻油和豬晶油混合，再與香葱肉碎拌勻，然後加精鹽和味精拌勻，由於新鮮香葱與鹽相遇後把其體內的水份排出，所以最好用篩子盛載，以便排水。

2　麵粉與生油混合，再徐徐拌入清水，揉搓至表面光滑的麵糰，蓋上保鮮紙或用濕布蓋好，醒發20分鐘，然後擀成長方形薄片，撒上乾麵粉，由外向裏捲成柱形，下劑子，並在橫切面按扁擀開。

3　將麵糰分成50克的劑子，按扁，擀開，放進30克餡料，再由內向外滾動成柱形，兩手各握一端，以左手為中心，並將右手圍繞轉圈，然後將尾端按壓在下，成圓餅生坯。

4　燒熱鍋，下油，油量要淹過煎餅2/3，燒至7成熟，放下葱油餅，煎至兩面金黃色，即可。

葱花大餅 /
福字餅

材料

- 芝麻適量，黏餅面用

麵糰
- 麵粉600克，砂糖1湯匙
- 依士1錢（4克，約1/2湯匙）
- 清水（北方人稱常溫水）350毫升

餡料
- 椒鹽1茶匙
- 白芝麻100克
- 花生油4湯匙
- 葱花200克
- 雞蛋漿50毫升

做法

1 將麵粉、依士粉、花生油拌勻，再分幾次加入常溫水開麵，揉至表面光滑，用保鮮紙或用濕布蓋好，醒發10分鐘。

2 在麵板或工作枱上撒上乾麵粉，放上已醒發好的麵糰，再次揉搓至表面光滑。

3 用擀麵杖將麵糰擀成長方形薄片，掃上油，撒上椒鹽，再撒上一層薄乾麵粉，最後撒上葱花，由外向內捲起成柱形。

4 將麵糰按所需大小出劑子，兩端向下弄摺，與底麵黏合，按扁，擀圓，刷上雞蛋漿，兩面均沾上白芝麻。

5 燒熱煎鍋，將兩面烙黃後加入清水100毫升，待水燒乾後再淋上油，煎至金黃色即可。

大廚提示

- 發麵餅的餅身厚，中間不易熟，所以要用水煎法處理，方可做出鬆軟又香口的特色。
- 做烙餅的麵糰不宜發得太大，故醒麵時約用10分鐘就可以了。
- 福字餅的做法一樣，只是不沾芝麻而改用印模烙字吧了！如沾了芝麻，即發麵芝麻香餅、上海家常餅或發麵大餅。

黃金大餅

材料

- 蛋漿適量，掃面用

麵糰

- 麵粉600克
- 依士粉1茶匙
- 泡打粉1茶匙
- 精鹽1/2茶匙
- 砂糖1湯匙
- 生油2湯匙
- 清水300毫升，後下

餡料

- 蘿蔔絲400克
- 精鹽2茶匙，醃蘿蔔用
- 生油2湯匙，炒蘿蔔用
- 葱花200克
- 火腿茸50克
- 白芝麻100克

調味

- 鹽7-8克
- 雞粉1湯匙
- 砂糖2湯匙
- 清水200毫升
- 生粉適量

做法

1 把蘿蔔絲與鹽撈勻，醃5分鐘，擠去水份。

2 燒熱鍋，下油，放入蘿蔔絲翻炒約3分鐘，下鹽、雞粉和糖炒勻，加清水和生粉勾芡，上碟，撒上葱花和火腿茸，放進雪櫃備用。

3 將麵糰材料（除清水外）混和，拌入清水，搓成麵糰，蓋上濕布，醒發1小時，再次揉搓至表面光滑。

4 將麵糰分出重100-150克的劑子，按扁擀圓，然後取一塊麵皮上放60-80克餡料，預留週邊，用水塗抹，再蓋上另一塊麵皮，壓緊。

5 取一隻碗反轉套住麵糰，切去多餘餅邊，掃上蛋漿，黏滿白芝麻，上鍋蒸10分鐘，燒熱油放下蒸餅，炸至金黃色即可。

圈圈餅
（草帽餅）

京醬肉絲

材料

- 豬瘦肉500克
- 京葱白100克，去皮，洗淨，切2吋長的幼絲

調味

- 生抽1湯匙
- 生粉2湯匙
- 麻油1湯匙
- 填鴨醬或甜麵醬1湯匙
- 豆腐皮4吋 × 4吋8塊（或單餅4張）

做法

1. 豬瘦肉洗淨，切絲，用生抽抓拌，靜置10分鐘，再加入生粉，一定用手抓均勻。
2. 燒熱油鍋，下1000毫升生油，油溫升至80℃左右，放入肉絲，用手勺推散肉絲，隨油溫的升高將肉絲炸至外焦內嫩，撈出，瀝去多餘的油。
3. 另起油鍋，煸炒甜麵醬，倒入炸好的肉絲翻勻，出鍋，上碟。
4. 擺上葱絲，伴豆腐皮或單餅就可以了。

材料

麵糰

- 冷水麵糰500克（請參閱第17頁）
- 燙麵糰150克（請參閱第17頁）
- 生油2湯匙

調味

- 麻油適量
- 椒鹽適量
- 麵粉適量

a　b　c　d
e　f　g　h
i　j　k　l

做法

1. 將冷水麵糰、燙麵糰和生油混合，揉搓至光滑柔軟的麵糰。
2. 將麵糰分成四等份，取其中一份在麵板上擀成長方形薄片，掃上麻油，撒上椒鹽，再撒薄薄一層麵粉。
3. 橫放麵皮，如摺紙扇一樣摺起，拉長，兩手各持一頭，以左手為圓心，繞成圓圈。
4. 燒熱鍋，下油，燒熱後放下餅，煎至兩面金黃，放在麵板上，蓋上濕布，用刀輕輕在餅身拍平。

叉子燒餅

肉碎炒玉豆

材料

- 絞碎豬瘦肉300克
- 毛豆600克，去皮留豆，葱花200克

调味

- 生抽2湯匙
- 味精1茶匙
- 麻油1湯匙，後下

做法

1. 起鍋燒水約1碗半，加八角1粒和精鹽1茶匙，下豆煮5分鐘，出鍋，瀝去水份。
2. 另起油鍋以50克葱花爆香，把肉碎下鍋，翻炒，加生抽2湯匙，續炒至熟。
3. 加入已煮好的毛豆和味精炒均勻，淋上麻油，撒上餘下葱花，出鍋即可。

材料

- 蛋漿適量，掃面用
- 芝麻適量，沾面用

麵糰

- 麵粉600克
- 生油2湯匙
- 清水350毫升

麵油酥

- 麵粉80克
- 生油50毫升

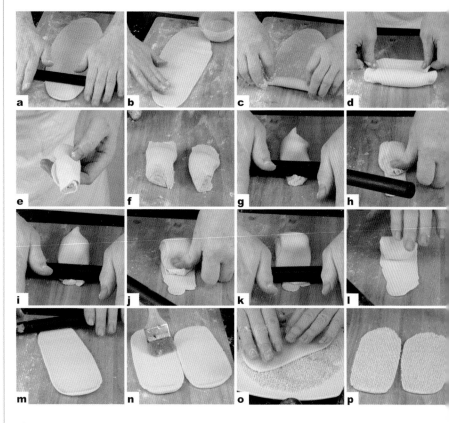

做法

1. 將麵粉、生油和清水混合，搓成麵糰，醒發30分鐘，然後把麵油酥材料混合，拌成糊狀。
2. 將已醒好的麵糰擀成長方形薄片，塗抹上麵油酥，由外向內捲起成柱形，以橫向捲最好，下劑子。
3. 從劑子橫向1/3的位置，向前擀開呈斜坡似，向前推捲，然後轉方向接口，向下約1/4的位置，再向前推擀成斜坡，摺起，換位擀另一端摺起見底。
4. 待所有麵糰都擀完，由第一個起將生坯橫放用手，壓扁，由中間起向前輕輕擀一下，轉換位置向前後擀開，刷上蛋漿，沾滿芝麻，預熱240℃烤箱烤10分鐘即可。

羊肉湯
泡饃

Chinese Pita Bread soaked in Lamb Soup

大廚提示

- 饃餅用了做「杠子麵」的製法，製成的饃餅，可置常溫下3個月不會變質。
- 羊肉湯泡饃多在冬季享用，只要加入1湯匙紅椒，既能禦寒又可飽肚。

材料

- 發酵麵糰400克（請參閱第15頁）
- 麵粉200克

羊肉湯

- 羊腿肉（羊肉片）900克
- 津白半棵，切絲
- 花椒適量
- 京葱適量
- 清水800毫升
- 香菜（即芫荽）5根

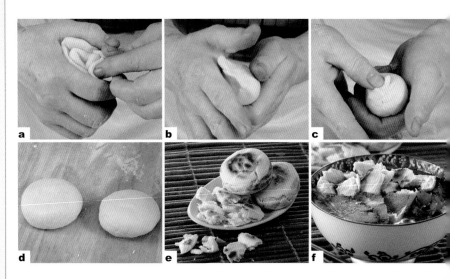

做法

1 將發酵麵糰加上麵粉搓勻，製成麵糰，分成5等份，壓扁。

2 先把麵糰放在餅鍋中烙至兩面金黃，再轉放焗爐120℃，烘焗15分鐘。

3 花椒和京葱稍炒，加入羊肉炒香，加入清水燒滾。

4 再加進津白絲，滾10分鐘。

5 洗淨香菜，切段，灑進羊肉湯裏。

6 食用時將饃撕成約大姆指的指甲狀大小，蘸湯食用即可。

南瓜豬肉餡餅

Pumpkin and Pork Stuffed Pan-fried Breads

材料

麵糰

- 麵粉600克
- 依士1茶匙
- 砂糖1湯匙
- 生油1湯匙
- 雞蛋1個
- 溫水350毫升

餡料

- 豬肉餡300克（請參閱第12頁）
- 南瓜600克

做法

1 南瓜去皮和挖去瓜瓤，洗淨，用磨絲器刨絲，再用刀攔（北方術語，南方術語是切）幾下，然後與豬肉餡混合，拌勻備用。

2 將麵粉、依士、砂糖、生油和雞蛋混合，搓勻，分幾次加入清水繼續搓揉，直至表面光滑的麵糰，蓋上濕布，置室溫下發酵10分鐘。

3 將麵糰下劑子，每糰50克，按扁，擀成圓皮，放進40克餡料，收起圓皮的邊，包成包子形狀，封口向下，壓扁，便成餡餅。

4 燒熱鍋，下油，放下餡餅煎至金黃，約2-3分鐘後反轉餡餅，再把另一邊煎黃即可。

大廚提示

- 因為煎餡餅時不加清水，所以調製麵糰時宜軟不宜硬。煎時，不要火太大，以免煎焦餡餅或導致外熟內生的狀況。
- 南瓜不宜切得太碎，否則影響口感。

芹菜豬肉餡餅

Stuffed Celery and Pork Pan-fried Breads

材料

麵糰	油酥	餡料
● 麵粉400克	● 麵粉200克	● 豬肉餡300克
● 生油1湯匙	● 生油100毫升	（請參閱第12頁）
● 清水220毫升		● 芹菜400克

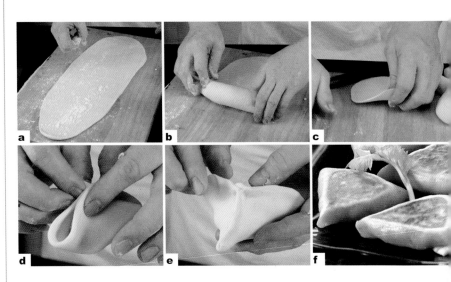

做法

1 將芹菜切碎，再與豬肉餡混合均勻，備用。

2 把麵糰材料拌勻，搓成軟滑麵糰，蓋上濕布，醒發15分鐘。

3 將油酥材料全部混合，攪拌成糊狀，備用。

4 將麵糰放在麵板上，揉搓至光滑，擀成長方形薄片，再塗上油酥糊，由外向內捲成柱形，下劑子，每劑子50克，按扁，按大，放進40克餡料，用虎口（拇指與食指的環形圈位）收緊封口，去掉多餘麵糰，按扁。

5 燒熱煎鍋，加油，放入餡餅，煎至兩面金黃即可。

翠瓜‧鮮蝦仁‧雞蛋餡餅

Stuffed Zucchini and Shrimp Pan-fried Beards

材料

麵糰	餡料	調味
● 麵粉600克	● 翠玉瓜3個	● 精鹽1茶匙
● 生油1湯匙	● 京葱1根	● 鮑魚汁2湯匙
● 常溫水350毫升	● 鮮蝦仁200克	● 生油4湯匙
	● 鹽1茶匙	● 麻油1湯匙
	● 雞蛋4隻，炒熟撕碎	

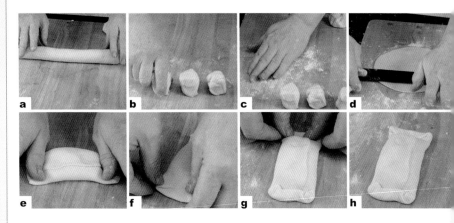

做法

1 洗淨翠玉瓜，擦絲（即用刨絲器刨絲），撒鹽拌勻，醃3分鐘後裝入煲湯布袋，壓去多餘水份，備用。

2 京葱去外皮，洗淨，切碎；鮮蝦仁洗淨，如太大隻要切小粒。

3 混合所有餡材料，加進鹽、鮑魚汁和生油先拌勻，再淋1湯匙麻油，放入雪櫃備用。

4 把麵粉與生油拌勻，再徐徐加入常溫水攪拌成絮狀，再搓揉成表面光滑麵糰，蓋上濕布，醒發20分鐘。

5 將麵糰下劑子，每劑子重約50克，按扁，擀成圓皮，先放餡料40克，前後對摺再封兩邊，即成餡餅生坯。

6 用中火燒熱煎鍋，放入盒子生坯，蓋上鍋蓋1分鐘，翻轉，再蓋鍋蓋直至熟透。

大廚提示

● 蔬菜焯水前於滾水中放入10克砂糖，這樣焯出的菜能保持鮮綠色。

● 綠豆粥的材料：大米200克，綠豆300克，清水1800毫升，全放在煲內煮稠，與餡餅同吃，十分滋味。

韭菜素盒子

材料

- 麵粉600克
- 生油1湯匙
- 常溫水350毫升

餡料

- 韭菜500克，切碎
- 生油50毫升
- 鮮蝦仁200克，炒熟
- 雞蛋4隻，炒熟
- 油條2件，切碎

餡調味

- 鹽1/2茶匙
- 雞粉1湯匙
- 麻油2湯匙

大廚提示

- 烙餡餅最好是多菜少肉，因為麵皮薄餡大，又淨肉餡就太肥了。
- 如以芹菜、四季豆、菠菜、小棠菜、白菜仔製餡，一定要焯水。

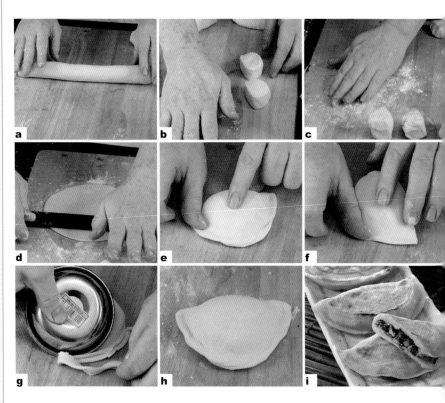

做法

1 韭菜碎與生油拌勻，並將餘下餡料和調味混合，拌勻，備用。

2 將生油和麵粉混合，一邊搓揉；一邊分3次加入溫水，搓至麵糰表面光滑，蓋上濕布，醒發20分鐘。

3 取出麵糰，搓成條，下劑子，每糰50克，按扁，擀成長橢圓形。

4 將麵皮置於左手掌心，放上40克餡料，用右手托起麵皮下垂部份，覆蓋在左手掌上，捏合圓邊，置於麵板，以一隻碗反轉切去多出的麵皮邊。

5 燒熱鍋，不用下油，用文火慢烙，直至麵皮上呈現斑點啡色即可。

誰不理
包子

Tianjin Buns Left Alone

典故

包子與水餃的餡料不同之處，前者可用熟肉、大塊肉、流沙和豆沙作餡；後者則不可以。所以包子品種真的很多，尤其是天津包子。其中有一種叫"狗不理"的天津包子，又叫"誰不理"，很有名氣，據說當中有段典故：有天有個包子店的老闆，因為沒甚麼客人，便站到門口開望，正好一位相識朋友經過，見他東張西望的，便笑問道：老闆生意可好嗎？老闆看看這位朋友，搖搖頭沒出聲，作了一個請進的手式，滑稽好笑！朋友搖搖頭，開玩笑的說：老闆你的包子皮大餡小，餵狗狗都不理，我不吃。就要走，老闆一聽"狗不理"便興奮了，說："狗不理"好啊，就"狗不理"吧。請進來坐吧。朋友心想只能"理"了，吃完包子，老闆沒有收他的錢，還謝了他。老闆從此改良了包子餡，同時改名為"狗不理包子舖"，至今生意都很紅火。

材料

麵糰

- 麵粉600克
- 依士1茶匙
- 泡打粉2茶匙
- 砂糖1湯匙
- 生油2湯匙
- 清水350毫升

餡料

- 豬上肉500克，瘦多肥少
- 淨豬瘦肉150克
- 京葱100克

餡調味

- 生抽1湯匙
- 麻油50毫升
- 甜麵醬25克
- 精鹽1茶匙
- 砂糖1湯匙
- 味粉1湯匙
- 火腿上湯300毫升
- 花生油150毫升
- 料酒2湯匙

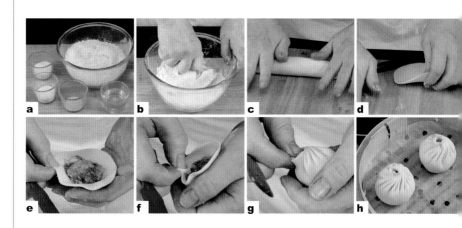

a b c d e f g h

做法

餡料

1 洗淨豬上肉，剁碎；洗淨豬瘦肉，切成花生粒大小，用生抽醃10分鐘，炒熟備用；洗淨京葱，切碎，加入麻油，拌勻備用。

2 在肉碎裏加上鹽、糖、味粉、料酒、花生油和上湯，順方向攪打至黏稠。

3 加上所有湯汁，再攪打至黏稠，拌入甜麵醬、葱碎和炒熟的肉丁，攪拌均勻入雪櫃備用。

麵皮

1 混合麵粉、依士、泡打粉、砂糖和生油，邊攪拌邊加水，直至加完水後，將粉料搓成大致光滑的麵糰，蓋上濕布，醒發1小時。

2 再將麵糰揉搓至表面光滑，蓋上濕布，10分鐘後搓成條，分成35克小糰，按扁，擀成中間厚週邊薄的圓片。

3 在麵皮裏加入35克餡料，用拇指與食指捏着麵皮的邊輕輕黏合，直至完成，靜放15分鐘。

4 在蒸鍋加水，在蒸籠鋪濕屜布或蠟紙，放進包子，至水滾蒸15分鐘即可。

大廚提示

- 「誰不理包子」和「狗不理包子」很相似，只是做法新穎，大同小異而已。

- 天津包子與狗不理如出一轍，只是它是多了蔬菜作餡料吧！餡料：淨豬肉餡300克、京葱100克、津白900至1200克、鹽2茶匙、鮮蝦仁200克、鮑魚汁1湯匙、麻油1湯匙攪勻。

生煎包

材料

- 清水250毫升，煎包用
- 麻油1湯匙，後下

麵糰

- 麵粉600克
- 依士1/2茶匙
- 砂糖1湯匙
- 生油2湯匙
- 生粉50克
- 清水330毫升，後下

餡料

- 豬上肉600克，絞碎
- 京葱50克，切碎
- 豬皮凍200克，捏碎後下
  （請參閱第73頁）

餡調味

- 精鹽1茶匙，砂糖1湯匙
- 味粉1湯匙
- 火腿上湯100毫升
- 花生油150毫升
- 料酒3湯匙

大廚提示

- 生煎包不用加泡打粉，因為泡打粉的特質很敏感，當與清水接觸後容易起泡。
- 麵糰加生粉有助防止湯汁外溢。
- 包好的生煎包，可用餡挑上水份，輕抹包子頂部，黏上葱碎或黑芝麻做裝飾。
- 好吃的生煎包的包底一定酥脆，一咬包子就有湯汁流出，還含有讓人食慾大增的紹興酒香。

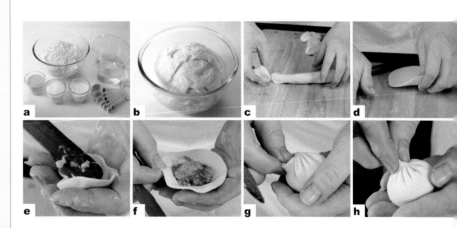

做法

1 將餡料和餡調味混合，攪打至黏稠，再拌入豬皮凍碎，盡量拌勻，然後淋上麻油，入雪櫃備用。

2 將麵糰材料混合，一邊攪拌；一邊徐徐加入清水330毫升，直至完成，搓揉成大致光滑的麵糰，蓋上濕布，醒發1小時。

3 再次將麵糰搓至光滑，蓋上濕布，醒發15分鐘，搓成長條，下劑子，每粒重35克，按扁，擀成中間厚週邊薄的圓片。

4 每一麵皮裹入35克餡料，用拇指與食指推摺麵皮的週邊，輕輕貼合，直至完成。

5 燒熱鍋，下油，油燒熱後放下生煎包，加入清水250毫升後蓋鍋，煮至水份完全揮發，淋上麻油，煎至包底金黃即可。

素菜包

材料

麵糰
- 麵粉600克
- 依士1茶匙
- 泡打粉2茶匙
- 砂糖1湯匙
- 生油2湯匙
- 清水350毫升

餡料
- 乾冬菇30克，
  浸發剁碎
- 葱1條
- 薑1塊
- 小棠菜4斤
  （2400克）

餡調味
- 精鹽1茶匙
  （約7-8克）
- 砂糖1湯匙
- 雞粉1湯匙
- 麻油1湯匙
- 油50毫升

a　b　c

d　e　f

g　h　i

做法

1　將葱和薑加油炒香，放入冬菇碎，一起炒到水份基本揮發，上碟，然後將薑葱和冬菇分開放。

2　洗淨小棠菜，汆水，沖凍水，剁碎，用布袋擠出水份，剩下約500-600克左右，加入餡調味和已炒冬菇碎混合拌勻，放入雪櫃。

3　混合麵粉、依士、泡打粉、砂糖和生油，然後一邊攪拌；一邊加入清水至完成，搓揉成大致光滑的麵糰，蓋上濕布，醒發1小時。

4　再次將麵糰揉搓至表面光滑，蓋上濕布，醒發10分鐘後下劑子，每粒重35克，按扁，擀成中間厚週邊薄的圓片。

5　每圓片加入35克餡料，用拇指與食指推摺週邊輕輕黏合，直至完成，靜放15分鐘。

6　在蒸鍋加水，在蒸籠鋪濕屜布或蠟紙，放進素菜色，蒸10分鐘即可（水滾計時）。

大廚提示
- 素菜包以蔬菜為主，聞起來清香可人，吃起來美味又健康。
- 蒸素菜包的時間不宜過長，超過10分鐘很容易變黃。

山東大包

材料

麵糰

- 麵粉900克
- 依士2茶匙
- 泡打粉2茶匙
- 砂糖1湯匙
- 生油2湯匙
- 清水500毫升

餡料

- 豬上肉150克，切丁炒熟
- 津白800克
- 京葱1條，切碎
- 香菜（芫荽）20克，切1厘米段
- 水發木耳150克，切指甲片狀
- 水發粉絲150克，切段
- 甜麵醬2湯匙

餡調味

- 鹽2茶匙
- 精鹽1/2茶匙
- 砂糖2茶匙
- 雞粉1湯匙
- 花生油50毫升
- 麻油1湯匙
- 五香粉1/4茶匙

大廚提示

- 山東大包以外形粗獷、餡料量足而味濃，咬下去富口感，十分獨特。

做法

1 洗淨津白，切成食指指甲大小，用2茶匙鹽醃10分鐘，用布袋擠去水份。

2 將所有餡料和餡調味混合，拌勻，放入雪櫃，備用。

3 將麵粉、依士、泡打粉、砂糖和生油混合，一邊加清水；一邊攪拌，直至完成，蓋上濕布，醒發1小時後，再度揉搓麵糰至表面光滑，醒發10分鐘。

4 將麵糰搓成長條，下劑子，每粒重50克，按扁，擀成長圓片，放進60克餡料，捏合麵皮，靜放15分鐘。

5 在蒸鍋加水，在蒸籠鋪濕屉布或蠟紙，放進山東大包，待水煮沸，以大火蒸15分鐘。

小籠包

包點和花卷

材料

麵糰
- "白菜牌"麵粉600克
- 生油2湯匙
- 清水300毫升

豬皮凍
- 生豬皮1斤（600克）
- 香葱4-5條
- 生薑1塊
- 料酒100毫升
- 清水4斤8兩（2700毫升）

餡料
- 豬上肉（瘦多肥少）1斤（600克），剁碎
- 京葱100克，切碎
- 豬皮凍400克，弄碎後下
- 麻油1湯匙，後下

餡調味
- 精鹽1茶匙
- 砂糖2湯匙
- 味粉1湯匙
- 花生油100毫升
- 料酒2湯匙
- 老抽1/2湯匙

a　b　c　d

e　f　g　h

做法

1　洗淨豬皮，汆水，剁碎，加入其餘材料，用大火煲5分鐘，待滾沸後調低火，續煲2.5小時，去渣，放涼，放入雪櫃雪20小時，便成豬皮凍。

2　將剁碎豬肉與餡調味混合，攪打至黏稠，加入豬皮凍碎，拌勻，淋上麻油，放入雪櫃備用。

3　將麵粉和生油混和，拌入300毫升清水，搓成麵糰，蓋上濕布，醒發20分鐘，搓至麵糰光滑。

4　將麵糰搓成長條，下劑子，每粒重16克，按成中間厚週邊薄的麵皮，放進20克餡料，將麵皮由下而上地打摺捏合，直至完成。

5　在蒸鍋加水，在蒸籠鋪濕屜布或蠟紙，水燒滾後才放進小籠包，蒸10分鐘即可（水滾計時）。

大廚提示

- 小籠包做到皮薄多汁味美，包製和蒸製就是成功關鍵。首先，一定要燒滾水才放小籠包，否則蒸氣不足，皮凍會溶解，甚至會塌陷。再者，蒸的時間不宜過長，否則包內的蒸氣過大，皮會爆破，湯汁外溢。最後，如果屜布沒浸水，小籠包會掉底。

- 市面上很多店舖採用魚膠粉調製上湯，每15斤（9000克）湯水加1斤6兩（825克）魚膠粉。如果家庭自製，則可以用市售150毫升上湯溶解60克魚膠粉，再燒滾450毫升上湯，加入調好的魚膠粉，煮滾，放涼即可放入雪櫃存24小時。

壽桃
百子壽桃

材料

- 食用紅花粉色素適量，刷面用

托盤麵糰
- 麵粉200克
- 依士1/2茶匙
- 泡打粉1茶匙
- 砂糖1湯匙
- 生油約1湯匙
- 清水100毫升

小壽桃
- 麵粉500克
- 依士1茶匙
- 泡打粉2茶匙
- 砂糖1湯匙
- 生油1湯匙
- 清水250毫升

百子壽桃
- 麵粉300克
- 依士1/2茶匙
- 泡打粉1茶匙
- 砂糖1湯匙
- 生油1湯匙
- 清水130毫升

百子壽桃餡
- 豆沙30粒
- 巧克力（朱古力）30粒

做法

托盤麵糰

將麵粉、依士、泡打粉、砂糖和生油混合，再用100毫升清水拌勻，揉搓至表面光滑，擀成直徑9吋的托盤，蒸熟。

小壽桃

1 將麵粉、依士、泡打粉、砂糖和生油混和，再用250毫升清水拌勻，揉搓至表面光滑，再下劑子8粒，每粒25-30克。

2 用碗製成大桃坯具，包上保鮮紙，將剩餘麵糰取2/3份，擀圓，包在模具，成桃形，貼上桃葉，蒸熟備用。

百子壽桃

將麵粉、依士、泡打粉、砂糖和生油混合，再用清水拌勻，揉搓至表面光滑，下劑子60粒，每粒約10克，並分別包入豆沙和巧克力，製成百子壽桃，蒸熟，放涼備用。

組合

將百子壽桃分別刷上一層食用紅花粉色素，再取下大桃套，裝入60粒百子桃，扣在蓮花盤中，四周擺放小壽桃即可。

大廚提示
- 所有發麵製品可以改用超級市場出售的自發粉，關鍵是添加料的份量是否適中。

刀切饅頭

包點和花卷

蝦仁椰菜炒年糕

材料

- 鮮蝦仁 15-20隻
- 青椰菜1/3個，切大塊
- 韓式年糕600克
- 甘筍8-10片

調味

- 精鹽2/3茶匙
- 鮑魚汁1湯匙
- 砂糖1湯匙
- 麻油1茶匙，後下

做法

1 蝦仁去泥腸，抓少少生粉撈勻，靜置10分鐘。

2 起鍋燒水煮至大滾，放入年糕，用手勺推動推散，離火，5分鐘後撈出，瀝水。

3 起油鍋，先煸炒蝦仁，再放椰菜和甘筍片一起炒幾秒，然後把所有材料下鍋和調味，翻炒均勻，待菜脫生，淋上麻油即可。

大廚提示

如選用上海或寧波的散裝年糕，一定浸水不少於8小時，方可使用。

材料

- 麵粉600克
- 依士1茶匙
- 泡打粉2茶匙
- 砂糖1湯匙
- 生油1湯匙
- 清水300毫升

做法

1 將麵粉、依士、泡打粉、砂糖和生油混合，再分3次拌入清水，揉搓成麵糰，蓋上保鮮紙，醒發30分鐘。

2 將麵糰再次揉搓至光滑，靜置10分鐘。

3 在麵板上撒上一層麵粉，將麵糰擀成長方形片狀，撒上少許麵粉，由外向內緊緊捲成柱狀，捲至最後的邊緣部位，壓到柱形下邊。

4 將麵糰按要求大小切段，便成饅頭，醒發20分鐘。

5 在蒸鍋加水，在蒸籠鋪濕屜布或蠟紙，放進饅頭蒸15分鐘即可（水滾計時）。

葱花卷

材料

麵糰
- 麵粉600克
- 依士1茶匙
- 泡打粉2茶匙
- 砂糖1湯匙
- 生油1湯匙
- 清水300毫升

餡料
- 精鹽1/2茶匙
- 胡椒粉1茶匙
- 五香粉1/2茶匙
- 葱花50克

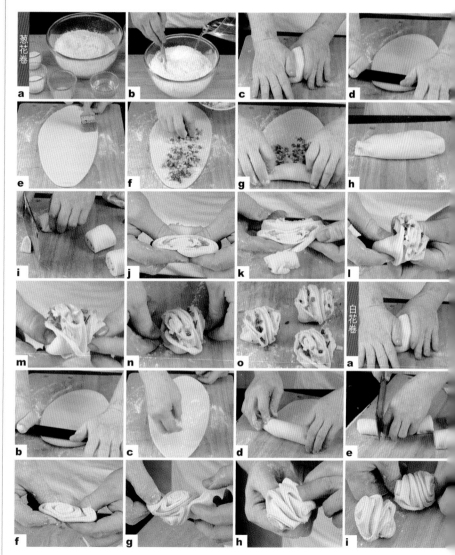

做法

1 將麵粉、依士、泡打粉、砂糖、生油和清水拌勻，揉搓，表面呈光滑的麵糰，醒發30分鐘，再揉搓至光滑，備用。

2 將鹽、胡椒粉和五香粉混合均勻，備用。在麵板上撒一層麵粉，然後將麵糰分成兩份，擀成長方形片狀，掃上生油，按序撒上椒鹽粉、薄麵粉和葱花。

3 將麵糰由外向內捲成柱狀，柱形不宜太粗，然後切段，用筷子從平面上橫向壓下，再用兩手拉長，且反方向扭180℃，兩手縱合，即封口，便成一個葱花卷。

4 在蒸鍋加水，在蒸籠鋪濕屜布或蠟紙，放進葱花卷蒸15分鐘即可（水滾計時）。

牛蹄花卷

八寶粥

材料

- 去核小紅棗20粒
- 佰荷20克
- 龍眼肉或葡萄乾 50克
- 清水2500毫升
- 砂糖適量

A料

- 小紅豆100克
- 栗子肉100克
- 肉豆50克
- 花豆80克

B料

- 蓮籽50克
- 糯米或大米100克

調味

- 生抽2湯匙
- 味精1茶匙
- 麻油1湯匙，後下

做法

1　A料先下鍋煮熟。

2　B料再下鍋，待B料熟後再放入其他料滾5分鐘，加糖即可。

材料

麵糰

- 麵粉600克
- 依士1茶匙
- 泡打粉2茶匙
- 砂糖1湯匙
- 生油1湯匙
- 清水300毫升

餡料

- 椒鹽1湯匙
- 麵粉100克
- 生油適量，刷麵糰用

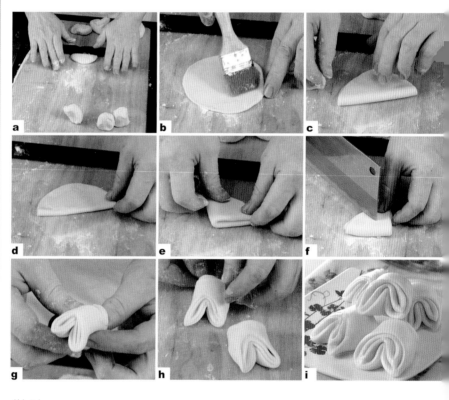

a　b　c
d　e　f
g　h　i

做法

1　將麵粉、依士、泡打粉、砂糖和生油混和，一邊搓揉；一邊加入清水，搓成表面光滑的麵糰，醒發30分鐘。

2　將麵粉和椒鹽拌勻，備用。

3　將已醒發的麵糰搓成長條，平均分成18份，按扁，擀圓，刷上生油，撒上少許椒鹽麵粉，再刷生油，對摺，並於尖角處切入2/3長度，兩邊向下，封貼好，醒發約15-20分鐘。

4　在蒸鍋加水，蒸籠鋪濕屜布或蠟紙，放進生花卷，蒸約10分鐘即可（水滾計時）。

銀絲花卷

材料

麵絲

- 麵粉300克
- 依士1/2茶匙
- 砂糖3湯匙
- 生油1湯匙
- 清水130毫升

麵糰

- 麵粉300克
- 依士1/2茶匙
- 泡打粉1茶匙
- 砂糖1湯匙
- 生油1湯匙
- 清水130毫升

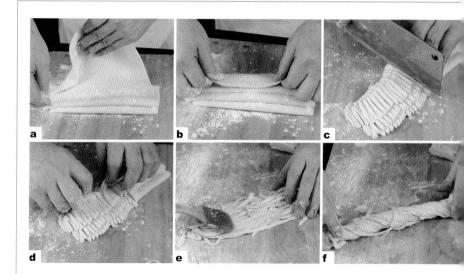

做法

麵絲

1 將麵粉、依士、砂糖和生油混合，用清水拌勻，揉搓至表面光滑。

2 把麵糰分成兩份，取其中一份放在麵板上，擀成長方形薄片，撒上麵粉，如摺扇子一樣，每摺一次都要撒一層麵粉。

3 用刀將麵糰切成條，為防止相互黏着，要適時鬆動，然後將所有麵條掃上油，平均分成12份，入雪櫃備用。

麵糰

1 將麵粉、依士、泡打粉、砂糖和生油混合，用清水拌勻，揉搓至表面光滑。

2 將麵糰平均分成12份，按扁，擀成長方形，摺上兩邊，放進麵絲，將兩邊多出的麵皮摺起，壓緊，再向前捲成長形，醒發45分鐘。

3 在蒸鍋加水，在蒸籠鋪濕屜布或蠟紙，放進花卷蒸10分鐘即可（水滾計時）。

銀絲卷

包點和花卷

材料

麵絲[1]

- 麵粉300克
- 依士1/2茶匙
- 砂糖3湯匙
- 生油1湯匙
- 清水130毫升

麵糰[2]

- 麵粉300克
- 依士1/2茶匙
- 泡打粉1茶匙
- 砂糖1湯匙
- 生油1湯匙
- 清水130毫升

[1] 商業做每回都會拉130至150條的麵絲，份量是6斤（3600克）麵粉，加依士6錢（23克）、豬晶油3兩（112克）、砂糖1斤8兩（900克），加清水大約3斤（1800毫升）。

[2] 至於商業做麵糰，份量是6斤（3600克）麵粉、依士8錢（30克）、泡打粉1兩5錢（57克）、砂糖12兩（450克）、豬晶油3兩（112克），加清水2斤4兩（1350克）。

大廚提示
- 拉麵時要快，尤其是氣溫高時，麵糰很快發起，便難於成絲。

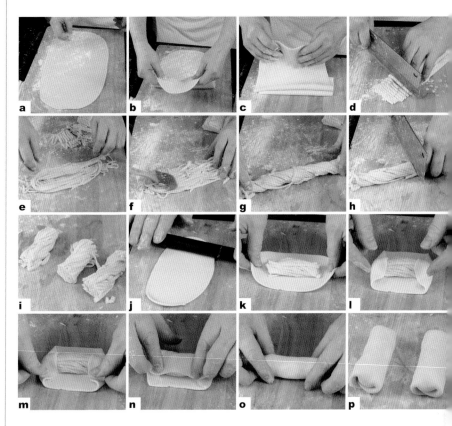

做法

麵絲

1 將麵粉、依士、砂糖和生油混合，加清水拌勻，揉搓至表面光滑的麵糰。

2 先將麵糰分成兩塊，然後在麵板上擀成長方形薄片，撒上麵粉，如摺扇子一樣摺起，每摺一次都要撒一層麵粉，然後用刀切成條，為防止相互黏着，要適時鬆動。

3 將所有麵條都掃上油，然後平均分成6份，放進冰箱備用。

麵糰

1 將麵粉、依士、泡打粉、砂糖和生油一起混合，然後加清水拌勻，揉搓至表面光滑，平均分成6份。

2 將麵糰按扁，擀成長圓形，直放麵絲，將兩邊摺上，然後向前捲成長形，醒發45分鐘。

3 在蒸鍋裏加水，在蒸籠鋪屜布或蠟紙，放進麵坯，蒸18分鐘（待水滾計時），可以油炸至表面金黃。

荷葉餅

醬牛肉

材料

- 牛小腿肉（腱子肉）
  4斤（2400克）
- 八角15克
- 桂皮20克
- 香葉5-6片
- 草果2-3個
- 黑胡椒2湯匙
- 葱薑40克
- 麵醬2-3湯匙

調味

- 老抽1湯匙
- 鹽1湯匙
- 砂糖2湯匙

做法

1. 燒滾水，把牛肉下鍋，焯一下，洗淨。
2. 另起油鍋，除牛肉外，把其餘材料下鍋炒香，加清水燒滾，放入牛肉和調味。
3. 轉小火煮到肉熟，出滷，攤凍，放入冰箱備用。

材料

- 發酵麵糰200克
  （請參閱第15頁）
- 生油適量
- 麵粉適量

餡料

- 生油適量
- 麵粉50克

a　b　c
d　e　f
g　h　i

做法

1. 將麵糰搓成條，分成10粒小麵糰。
2. 將麵糰按扁、擀成長圓形，刷上生油，撒少許麵粉，對摺4下。
3. 用刀割上斜紋，造型，醒發15分鐘。
4. 把荷葉餅坯放在蒸鍋上，用大火蒸7分鐘即可（水滾計時）。

大廚提示

- 醬牛肉、醬驢肉、醬羊肉或醬豬肉，這些都是北方不可少的食物，皆是滷製品，適合北方天氣乾燥，冬天時間長易存放滷味香純長年不衰。

豬腦花卷

木耳番茄雞蛋湯

材料

- 水發木耳6朵
- 紅番茄1斤（600克）
- 上湯1200毫升
- 雞蛋3隻，打散
- 香菜（芫荽）2根，切段

調味

- 鹽1茶匙
- 鮑魚汁2湯匙
- 生粉水3湯匙

做法

1 紅番茄以滾水焯一下，去皮，一開為六。
2 起鍋加1湯匙生油，炒腍番茄，加入木耳和上湯。
3 下鹽和鮑魚汁燒至大滾，加入生粉水，淋上雞蛋液，撒上香菜即可。

材料

- 發酵麵糰600克（請參閱第15頁）
- 生油適量
- 麵粉適量

餡料
- 生油適量
- 麵粉50-100克

a
b
c
d
e
f

做法

1 將發酵麵糰分3等份。
2 每份麵糰擀成10×15公分的長方形薄片，掃上生油，撒上些麵粉，摺4-5摺。
3 拉長麵糰，然後將3條互相纏繞，連接處在下，置室溫醒發15分鐘。
4 把生豬腦花卷放在蒸鍋上，用大火蒸10分鐘即可（水滾計時）。

大廚提示

- 番茄湯的變化很多，如改用鮮蝦燴鍋湯，其鮮味十足，另外加番薯塊、新鮮粟米或薯仔都有不同的感覺。

材料

- 發酵麵糰600克
  （請參閱第15頁）
- 生油適量
- 麵粉適量

餡料

- 生油適量
- 麵粉50-100克

香椿芽炒雞蛋

材料

- 香椿芽6-8根
- 雞蛋4隻，打散
- 京葱白適量，切碎

調味

- 精鹽1/3茶匙
- 味精1茶匙

做法

1 新鮮的香椿芽用鹽醃10分鐘，洗淨，切碎；要是用醃製的貨品，就要浸水5-10分鐘，洗淨，切碎。

2 將雞蛋液與切好的香椿芽加調味拌勻。

3 起鍋，加3湯匙生油，熱至7成熱，放入葱碎熗鍋，炒香，倒入所有材料翻炒至熟即可。

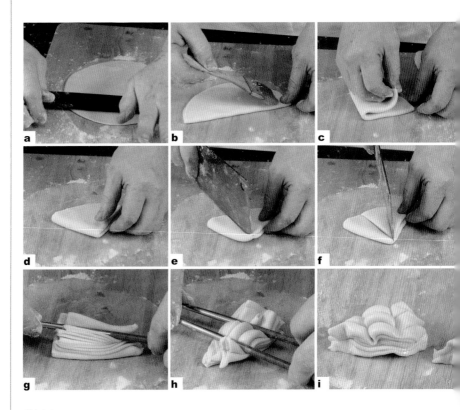

a　b　c
d　e　f
g　h　i

做法

1 將發酵麵糰分3等份。

2 每麵糰擀成20公分直徑的圓片，掃上生油，撒些麵粉，對摺，再刷油，再對摺。

3 用刀切成3份，疊合，用一隻筷子順向壓下，再用兩隻筷子於1/3處和2/3處橫向壓下。

4 用手封住麵糰兩端，醒發15分鐘。

5 把生蓮花卷放上蒸鍋，用大火蒸10分鐘即可（水滾計時）。

花素蒸餃

雜類

材料

麵糰

- 燙麵糰300克
  （請參閱第17頁）
- 冷水麵糰300克
  （請參閱第17頁）

餡料

- 白菜仔2斤（1200克）
- 小棠菜2斤（1200克）
- 水發木耳100克，
  切碎
- 水發冬菇100克，
  切碎
- 水發粉絲100克，
  切碎

餡調味

- 精鹽1茶匙
- 砂糖2湯匙
- 味粉1湯匙
- 薑葱油100毫升
- 鮑魚汁1湯匙
- 麻油2湯匙

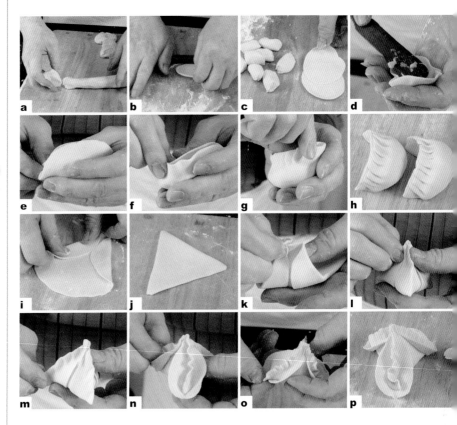

a　b　c　d
e　f　g　h
i　j　k　l
m　n　o　p

做法

1. 將白菜仔和小棠菜洗淨，焯滾，沖凍水，剁碎，放布袋擠去多餘水份，備用。

2. 將所有餡料混合，加入精鹽、砂糖、味粉、薑葱油和鮑魚汁，攪拌均勻，淋上麻油，放入雪櫃備用。

3. 將燙麵糰和冷水麵糰混合，揉搓成麵糰，一邊搓揉；一邊加入麵粉，使其更光滑。

4. 將麵糰下劑子，每粒重35克，按扁、擀圓皮，放進35克餡料，用拇指將麵皮向上推，推一下，以食指和拇指摺一下邊，直至整個餃都封好。

5. 在蒸鍋加水，在蒸籠鋪濕屜布或蠟紙，待水滾後，放進素餃蒸10分鐘即可。

大廚提示

- 蒸餃餡或鍋貼餡若選用生的蔬菜，那麼，麵皮比例應以1份燙麵糰對5份冷水麵糰；因為生的蔬菜煮熟後會縮小，導致皮大餡小，影響口感。建議包餃時，多放點餡料，以備它收縮後與麵皮比例相稱。有見及此，麵皮需要較大彈性，但是燙麵糰缺乏彈性，不可包入太多餡；所以要增加冷水麵糰的份量，增強彈性。

羊肉燒賣

材料

麵糰	餡料	餡調味
• 燙麵糰400克 （請參閱第17頁）	• 新鮮羊肉300克	• 麻油1湯匙
• 冷水麵200克 （請參閱第17頁）	• 熟糯米200克	• 生抽1湯匙
	• 鮮蝦20隻	• 味粉1/2湯匙
		• 花椒粉1/4湯匙

大廚提示

• 市面售賣的上海餛飩皮或水餃皮都可以作燒賣皮。

做法

1 新鮮羊肉洗淨，切粒；糯米蒸熟，趁熱拌入麻油；鮮蝦洗淨，去殼留尾。

2 將生抽和味粉放入羊肉粒裏拌勻，灑下花椒粉，攪至黏稠，然後加入熟糯米，拌勻，放於較暖的地方。

3 將燙麵糰和冷水麵混合，揉搓成表面光滑麵糰，下劑子，每粒重20克，按扁，擀成圓麵皮。

4 左手托麵皮，放入20克餡料和1隻鮮蝦，以虎口位由圓皮中心至週邊1/3處向上收，即成燒賣。

5 在蒸鍋加水，在蒸籠鋪濕屜布或蠟紙，放進燒賣蒸10分鐘即可（水滾計時）。

炸醬麵

炸醬材料

炸醬

- 豬肉碎500克
- 甜麵醬250克
- 水發木耳70克，粗剁
- 雞蛋1隻，打散

調味（炸醬）

- 味鹽1茶匙
- 砂糖1湯匙
- 清水/上湯50毫升

做法

燒熱鍋，加入500克肉碎生炒至變色，加入甜麵醬一起炒至麵醬溢出香氣，落調味煮1分鐘左右，加入木耳碎，再撒上雞蛋漿。

材料

手切麵條

- 麵粉400克
- 精鹽1/2茶匙
- 冷水180毫升
- 薯仔2個（約350克），切絲炒熟
- 豆角（四季豆）350克，切段炒熟
- 青瓜1條（約200克），切絲
- 甘筍200克，切絲焯水
- 香菜4-5根，切碎
- 蒜肉4粒

蒜泥調味

- 鹽1/4茶匙
- 白開水40毫升
- 醋1湯匙
- 麻油1茶匙

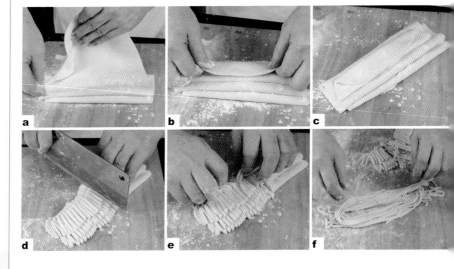

a　b　c
d　e　f

做法

1　蒜肉拍過，加鹽待片刻，搗成泥，再加白開水、醋和麻油拌勻。

2　將麵粉和鹽拌入冷水中，搓成麵糰，蓋上濕布，醒發30分鐘後。

3　將麵糰一分為二，搓至表面光滑，用大擀麵杖擀成長方形薄片，撒上麵粉，如摺扇子一樣，摺一次，撒一層麵粉，然後切成麵條，為防相互黏着，需要及時鬆動。

4　燒一鍋滾水，放下麵條，用筷子攪動着煮5-6分鐘，撈起分4份，按序放上蒜茸、青瓜絲、炒薯仔絲、甘筍絲、炒豆角絲，中間放入炸醬即可。

大廚提示

- 北方人做撈麵，會配很多菜如水發黃豆、炒花生瓣、炒蠶豆瓣、炒芽菜、炒韭菜、焯水菠菜、炒津白絲、拌香椿芽、番茄或打滷拌麵等等。

大滷麵

材料

麵條
- 麵粉600克
- 精鹽1/2茶匙
- 冷水250毫升

滷料
- 豬瘦肉150克
- 生抽1湯匙，醃豬瘦肉用
- 鮮蝦仁200克，切片
- 水發海參150克，切片
- 水發木耳80克，切片
- 水發黃花菜（即金針菜）50克，去根部
- 水發冬菇3朵，切片
- 清水250毫升
- 雞蛋2隻，打散後下
- 麻油1茶匙，後下

調味
- 老抽1/2湯匙
- 鮑魚汁1湯匙
- 精鹽7-8克
- 味精1茶匙

芡汁
- 生粉20克
- 清水3湯匙

做法

1 豬瘦肉洗淨，切絲，拌入生抽，備用。

2 將麵粉和精鹽拌入冷水中，搓揉成麵糰，蓋上濕布，醒發30分鐘。

3 將麵糰分成兩等份，揉搓至表面光滑，用大擀麵杖擀成長方形薄片，撒上麵粉，如摺扇子一樣，摺一次，撒一層麵粉，然後切成麵條，為防止互相黏着，適時鬆動。

4 燒熱鍋，放2湯匙油燒熱，放入瘦肉絲、蝦仁和冬菇片，炒香，加入清水，燒滾。

5 加入海參、木耳、黃花菜和調味，改小火煮10分鐘，勾芡，淋上雞蛋漿，再淋麻油即成。

6 另燒一鍋滾水，放下手切麵條，用筷子攪動，不能讓麵條黏在一起，煮5-6分鐘，撈起，分4碗，放上大滷料即成。

長壽麵

Longevity Noodles

材料

- 冷水麵糰260克
  （請參閱第17頁）
- 麻油適量，後下
- 香菜1根，切碎

湯料

- 瘦肉絲20克
- 番茄2個（一分為四）
- 水發木耳4朵
- 鮮蝦仁6隻
- 上湯250毫升
- 雞蛋1隻，打散後下

調味

- 鹽適量
- 鮑魚汁適量
- 生粉水2湯匙

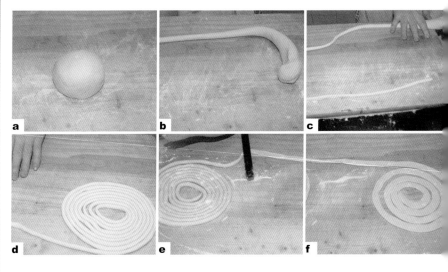

a　b　c
d　e　f

做法

1 將冷水麵糰搓成長條，擀扁，撒上麵粉，備用。

2 燒熱鍋，放瘦肉絲和蝦仁炒香，再加入番茄炒熟

3 放上湯和木耳煮滾，加調味拌勻，淋上雞蛋漿。

4 另備一個鍋，放麵條煮熟，置碗中。

5 將煮好的湯淋在麵條碗裏，下麻油，撒上香菜，一碗長壽麵就做成了。

大廚提示

- 北方人愛做這款家常麵，一碗只有一條長麵，寓意長壽和吉祥。

- 在酒樓設宴做生日，老闆必送之長壽麵有葷有素，千萬不能送餅或有餅的餸；因為在製餅過程中要左翻右翻，寓意不好。

貓耳朵
珍珠麵

材料

- 冷水麵糰500克
  （請參閱第17頁）
- 麵粉50克
- 麻油適量

三鮮湯料

- 瘦肉絲400克
- 水發木耳8-10朵
- 水發海參3條，切片
- 鮮蝦仁30隻
- 水發冬菇6朵
- 上湯1500毫升
- 雞蛋3隻，打散
- 唐生菜葉6片

調味

- 鹽適量
- 鮑魚汁適量
- 老抽適量

芡汁

- 生粉2湯匙
- 清水3湯匙

做法

1 把冷水麵糰取一半搓成如筷子般粗的麵條，再切成不足1公分的麵粒，用大拇指碾壓成小碗狀，為貓耳朵備用。

2 將另一半冷水麵糰擀成薄片，撒上麵粉，摺三摺，用刀切成條，撒上麵粉，切成小粒，為珍珠麵。

3 用熱鍋炒香肉絲，加入上湯，燒滾，下調味，加入木耳、海參、蝦仁和冬菇，勾芡，淋上雞蛋漿。

4 另備一個鍋燒水，放入貓耳朵和珍珠麵分別煮熟。

5 將生菜放於煮過麵的水裏焯一下，與貓耳朵和珍珠麵放進湯裏，便是貓耳朵珍珠三鮮湯。

大廚提示

- 貓耳朵、珍珠麵與刀削麵，如出一轍。

羊肉
燴湯麵

Mutton Soup Noodles flavoured with Sichuan Pepper Oil

材料

- 冷水麵糰1200克
  （請參閱第17頁）

羊肉燴湯

- 羊腿肉600克，切絲
- 津白1000克，切粗絲
- 水發粉絲400克
- 香菜200克，切段
- 甜麵醬4湯匙
- 指天椒2粒，切碎
- 花椒15克
- 上湯1000毫升

調味

- 鹽2茶匙
- 醋2湯匙

大廚提示

- 「燴」的意思是用油將花椒炒香，然後用來爆炒其他食材。羊肉燴湯麵就用了這種做法。

- 南方做菜不會用此烹調技法，所以他們的菜牌不會有燴芹菜、燴黃瓜（青瓜）、燴洋白菜（椰菜）或燴土豆絲（薯仔絲）等菜式。

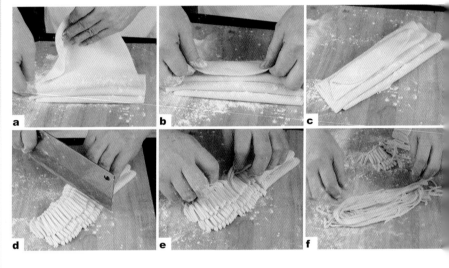

做法

1 冷水麵糰擀成薄片，切成條。

2 煮熟手擀麵，沖凍水，備用。

3 燒熱鍋，炒香花椒，濾去花椒粒。

4 放羊肉絲和甜麵醬，炒香，再放津白絲，略炒，加入上湯煮滾。

5 放進調味，加入備好的麵條，撒上香菜，便是羊肉燴湯麵。

臊子麵

材料

特細手擀麵
- 冷水麵糰600克
  （請參閱第17頁）
- 麵粉30克

湯料
- 豆腐1塊
- 羊肉絲400克
- 水發木耳6朵
- 水發黃花菜50克
- 香菜4根
- 水發冬菇6朵
- 紅椒粉20克
- 花椒20克
- 酸菜碎40克
- 雞蛋2隻
- 上湯1500毫升

調味
- 鹽2茶匙
- 味精2茶匙
- 老抽1/2湯匙，
  調色用
- 生粉水3湯匙
- 麻油1/2湯匙，
  後下

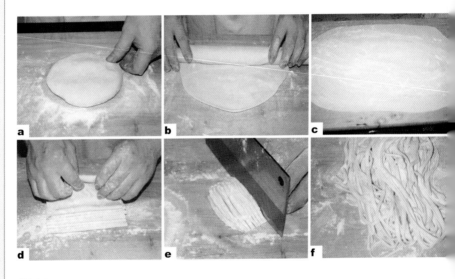

做法

1 將冷水麵糰盡量擀薄，切細，撒上麵粉備用。

2 將豆腐切成1公分大小；香菜洗淨，切成段；水發冬菇切片，備用。

3 燒滾水，放入手擀麵，單方向攪動至熟，撈出，過凍水，備用。

4 燒熱鍋，爆香花椒、羊肉絲和紅椒粉，加入上湯，煮滾，放豆腐塊、冬菇、木耳和酸菜，煮熟。

5 放鹽和味精調味，下老抽調色，煮滾，加入生粉水，再淋上雞蛋漿，滴落麻油，撒上香菜，放入麵條，輕攪動小滾即熄火。

大廚提示

- 純北方的一種湯麵，具有酸、鹹、辣、麻、香的特性。

- 這款麵不是每家媳婦能做好。因為它的精髓在於肉肉碎多而香脆；麵條細如鬆絲，入口即溶。至於臊子就如同炸肉醬，昔日只有富貴人家才能享用，因為肉末用量很多，才能做出效果，小戶人家花不起錢買肉，所以會添加水煮肉，變成軟炸醬，不合標準。

- 我在誤打誤撞下，因不愛市面的辣椒粉，改用潮州辣椒醬，加上多配了台灣蕎頭佐麵，效果很好。

北京炒麵

雜類

手擀麵

材料

- 麵粉800克
- 清水350毫升
- 鹽1茶匙

做法

1 與冷水麵糰相似（請參閱第17頁）。
2 置一旁醒約20分鐘，然後用大擀麵杖擀成薄片，摺疊後再切成幼絲。

材料

- 手擀麵1200克

配料

- 韭黃400克
- 豬肉絲400克
- 鮮蝦仁200克
- 上湯100毫升

調味

- 鹽1.5茶匙
- 鮑魚汁1湯匙

做法

1 手擀麵放入滾水煮熟，撈出，過凍水備用。

2 韭黃洗淨，切成1吋長的小段。

3 燒熱鍋，將韭黃、肉絲和蝦仁炒香。

4 加入上湯，再放進麵條，炒勻，下調味炒勻，上碟。

大廚提示

- 北方人開麵時，加少許鹽，有助麵皮起筋，煮好的麵條富彈力和韌度。

八寶
糯米飯

Eight Treasures Glutinous Rice

材料

- 白糯米600克
- 去皮熟栗子10粒
- 圓片菠蘿3片
- 紅棗茸500克
- 紅棗40粒（水煮過）
- 葡萄乾200克
- 果單皮1粒
- 蘆葦葉10片
- 熟花豆35粒
- 砂糖200克

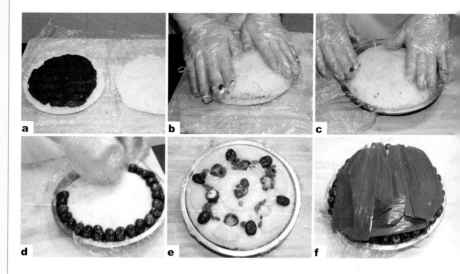

做法

1 將白糯米泡水6小時，隔水蒸20分鐘。

2 取350克熟糯米，拍製成直徑20公分的圓餅；紅棗茸也拍製成直徑20公分的圓餅，放於糯米圓餅上。

3 用剩餘的熟糯米同樣製成圓餅，蓋在紅棗餅上，將1片圓菠蘿片放在中間。

4 將餘下2片菠蘿片疊起，1開4，背向上，分鑲入糯米中。

5 在菠蘿片之間各放1粒栗子和紅棗，然後在圓菠蘿片中間放1粒栗子。

6 將果皮剪出公仔，放於空白處，再梅花間竹地圍上花豆和紅棗，最後包上保鮮膜，扣上大小相約的深盤，然後整體翻轉，蓋上蘆葦葉，放鍋裏蒸30分鐘。

7 將砂糖加適量水煮至濃稠，食用前淋在糯米飯上即可。

自製豆漿

材料
- 精選黃豆300克
- 清水4斤（2400克）

做法

1 黃豆洗淨，用800毫升清水浸不少於8小時，隔夜最好，已浸泡的黃豆約有900克之多。

2 取1800毫升清水用來打爛黃豆；另600毫升清水放入湯煲燒滾，浸泡黃豆的豆水，最好用來澆花，不宜食用。

3 打豆時，不要心急，可分3-4次打成豆漿，倒入布袋，濾出豆漿，同時用手擠壓，使豆漿快速流出。

4 將豆漿倒入滾水中一起煲，當豆漿溫度升至70℃左右，會產生大量泡沫，這泡沫是豆漿的精華，所以要不停的攪動，同時地將火轉小，這時的豆漿會滾起泡沫，隨着泡沫漸減至完全消失，待滾5-10分鐘才是最佳豆漿。記住沒有滾透的豆漿會產生皂素，食用後可能會腹瀉。

大廚提示
- 自製豆漿變化多端，配合其他材料便變成黑芝麻豆漿、亞麻籽豆漿、花生豆漿等等，當豆漿煮至八成滾，加入適量的石膏粉，15分鐘後倒入放有紗布的模具中，壓上重物1小時後即成豆腐。

回鍋肉

材料
- 椰菜1/3個，切大塊
- 青紅圓椒各1個，切大塊
- 洋葱1/4個，切大塊
- 大蒜苗，切1吋段．
- 五花肉8片（約4兩，150克），八成熟
- 五香豆腐乾1塊，斜切6片。

調味
- 豆瓣醬1湯匙
- 甜麵醬2湯匙
- 蒜頭碎1湯匙
- 精鹽1/2茶匙
- 味精1茶匙

做法

1 燒一鍋熱油至約160℃，將所有材料下鍋，即撈起，再用熱水沖去表面浮油，瀝水。

2 另起油鍋，放蒜碎熗鍋，下調味，把材料放入翻炒均勻，淋上麻油出鍋即可。

大廚提示
- 材料過熱油，是為了保留各蔬菜中的維生素，不被破壞。如果過熱水，時間較長，蔬菜中的維生素會被水帶走。
- 如果在家中製作，可先用熱油鍋炒軟材料，出鍋放在篩子中瀝水，另起油鍋再炒，添加1湯匙的生粉水勾芡就可以。

Pork Filling
豬肉餡 (P12)

Time: 10 mins   Made: 1200gm

Ingredients

600gm Pork front leg, 200gm Scallions, 20gm Gingers

Seasonings

100ml Soy sauce, 1 tablespoon Dark soy sauce, 50ml Cooking wine, 1.5 teaspoon (9gm) Fine salt, 2 tablespoon (24gm) Granulated sugar, 10gm MSG, 270ml Cold water, 2 tablespoon Sesame oil

Procedures

1. Rinse pork thoroughly and then mince. Put in a mixing bowl; add cooking wine, soy sauce and dark soy sauce. Stir and mix vigorously in one direction until pale surface. Add salt, sugar, MSG and 67.5ml cold water. Stir and mix until sticky and elastic texture. Add half of the remaining portion of water, repeat stirring procedures.
2. Pour in the remaining amount of water and repeat same procedures. Add scallions, gingers and sesame oil, stir and combine well.

Beef Filling
牛肉餡 (P12)

Time: 10 mins   Made: 1200gm

Ingredients

300gm Fresh beef, minced, 300gm Fat beef, minced, 150gm Scallions, skin and chopped

Seasonings

50ml Cooking wine, 1.5 teaspoon (9gm) Fine salt, 2 tablespoon (24gm) Granulated sugar, 10gm MSG, 1/2 teaspoon 5-spice powder, 150ml Oil, 2 tablespoon Sesame Oil, 150ml Cold water

Procedures

1. Place minced fat and beef inside mixing bowl. Add salt, sugar, MSG and oil. Stir and mix vigorously in one direction until sticky and elastic.
2. In between pour in mixing bowl 150ml of water (divide into 3 times), repeat stirring and mixing procedures after each time water is added. AT the end, sprinkle chopped scallions and sesame oil. Mix well.

Mutton Filling
羊肉餡 (P13)

Time: 10 mins   Made: 1200gm

Ingredients

300gm Mutton leg, 300gm Mutton belly, 200gm Scallions, 5gm Dried Sichuan pepper powder

Seasonings

200ml Oil, 1.5 teaspoon (9mg) Fine salt, 2 tablespoon (24gm) Granulated sugar, 10gm MSG, 2 tablespoon Sesame oil, 2 tablespoon cooking wine, 200ml Cold water

Procedures

1. Rinse thoroughly both kinds of mutton separately. Mince. Peel scallions, rinse and mince.
2. Combine minced mutton with salt, sugar, MSG, cooking wine and oil in a mixing bowl. Sprinkle in sieved Sichuan pepper powder. Stir and mix vigorously in one direction until sticky and elastic.
3. Add 1/3 portion of water, stir and mix vigorously in one direction until sticky and elastic. Repeat all steps until other 2 portions of water utilized. Add minced scallions and sesame oil, combine well.

Chicken Filling
雞肉餡 (P13)

Time: 10 mins   Made: 1000gm

Ingredients

300gm Chicken thigh meat, 300gm Chicken breast meat, 100gm Scallions (bulb only), 20gm Ginger, finely chopped

Seasonings

150ml Soy sauce, 50ml Cooking wine, 1 teaspoon (6gm) Fine salt, 2 tablespoon (24gm) Granulated sugar, 1 tablespoon (15gm) Chicken powder, 100ml Cool water, 2 tablespoon Sesame oil, 2 Egg white

Procedures

1. Rinse and clean chicken meat. Mince or chop finely. Skin scallions, remove foliage, and chop bulb finely.
2. Put meat, salt, sugar, chicken powder, cooking wine, soy sauce and egg white in a mixing bowl. Add 50ml of water. Stir and mix vigorously in one direction until sticky and elastic.
3. Add another 50ml of water. Stir and mix vigorously in one direction until sticky and elastic. Merge with minced scallion bulb and sesame oil.

Fish Filling
魚肉餡 (P14)

Time: 10 mins   Made: 900gm

Ingredients

1 Mackerel (about 2 catties or 1200gm), 50ml Fresh gingers, 100ml Water, 100gm ice cubes, 6 Egg white, 2 teaspoon pepper

Seasonings

4 tablespoon (50gm) Corn starch, 1 tablespoon Fish sauce, 1 teaspoon (6gm) Salt, 1 tablespoon (12gm) Granulated sugar, 1 tablespoon (15ml) White vinegar, 100ml Oil, 100gm Jett brand lard, Jett brand lard: good quality of lard, manufactured from The Netherlands

Procedures

1. Put gingers, water and ice cubes in blender to make ginger juice.
2. Rinse mackerel thoroughly. Remove head, bones and skin. It should remain 550-700gm of flesh.
3. Mince fish meat by mincer for 2 times to ensure all thin bones are smash. Then put in all ingredients and seasonings, stir and mix in one direction until sticky and elastic.

Veggie Filling
素食餡 (P14)

Time: 10 mins   Made: 900gm

Ingredients

2 Old cucumbers (about 3.5 catties or 2100gm), 5 Scrambled eggs, chopped, 100gm Soaked black fungus, chopped, 200gm Fresh skinned shrimps, 100gm Scallions, rinsed and chopped, 150ml Oil, 1 teaspoon (6gm) Fine salt, 2 tablespoon (24gm) Granulated sugar, 10gm MSG, 2 tablespoons Sesame oil

Procedures

1. Skin and core old cucumbers. Rinse. Shred and then sprinkle 2 teaspoons, mix well. Marinate for 5min.
2. Squeeze and drain juice. It should get about 500-600gm of flesh
3. Place cucumbers on cutting board and apply a few cuts. Put in mixing bowl. Add in all ingredients and seasonings. Combine well.

Dough 麵糰

Fermented Dough
發酵麵糰 (P15)

Time: 1 hour   Made: 1000gm

Ingredients

600g flour, 1 tsp yeast, 2 tsp baking powder, 2 tbsp sugar, 2 tbsp oil, 300ml water

Method

1. Mix flour, yeast, baking powder, sugar and oil together.
2. Add in 300ml water by three times and keep on kneading to form smooth dough.
3. Cover the dough with cling wrap. Rest at room temperature for 50 minutes. Then knead for 10 minutes more to uncooked small dough. Let them ferment for further 20 minutes.
4. Put the fermented dough in a steamer. Cook for 20 minutes on high heat.

Demi-fermented Dough
半發酵麵糰 (P16)

Time: 10 mins   Made: 1000gm

Ingredients

600g flour, 1 tsp yeast, 1 tbsp sugar, 1 tbsp oil, 1 pc egg, 350ml water

Method

1. Put flour, yeast, sugar, oil and egg to mix well.
2. Adding water in the mixture to keep on stirring until flaky. Knead to smooth dough. Cover with wet cloth and rest for 10 minutes.
3. Tear the dough into small pieces. Press flat and roll to round sheet for use.

Cold Water Dough
冷水麵糰 (P17)

Time: 10 mins   Made: 800gm

Ingredients

600g flour, 300ml water

Method

1. Add water in flour by three times and stir well.
2. Knead to smooth dough. Cover with plastic wrap tightly. Rest at room temperature for 20 minutes. Ready for use.

Cooked Dough
燙麵糰 (P17)

Time: 10 mins   Made: 1000gm

Ingredients

600g flour, 2 tbsp oil, 500ml boiling water

Method

1. Mix oil and flour together. Pour boiling water into the mixture rapidly. Use a pair of chopsticks to stir in dry flour until combined.
2. When the mixture tends to be a dough, knead by hands. Beware of scalding. It is cooked dough. Make two or three splits on the dough with a knife to release hot air.

Dumplings and Pan-fried Dumplings
餃子與鍋貼類

Shiitake and Beef Dumplings
香菇牛肉水餃 (P18)

Time: 8-10 mins   Serve: 4-6

Ingredients

Cold dough

600gm Low protein flour, 280ml Water, 1/4 teaspoon Fine salt

Stuffing

400gm Beef filling (please refer to page 12), 1 Onion, skined and chopped, 200 gm Soaked shiitake, diced, 2 tablespoon Sesame oil, added lastly

Procedures

1. Combine all filling ingredients and mix well. Add sesame oil and mix well. Set aside.
2. For making cold dough, please refer to page 17. Ferment for 15min.
3. Remove a portion of dough from mixing bowl. Roll it into a cylinder as thick as a thumb. Divide into evenly sized pieces, about 19-22gm per piece. Slightly stretch out each dough and using rolling pin, roll them flat and make edges thinner.
4. Put a wrapper on table, spoon filling onto the middle, Press a bit.
5. Fold the wrapper in half. Crimp the edges and pinch together to seal.
6. Bring water to a boil. Stir water to form swirls. Place dumplings inside and stir slightly a bit to avoid them getting burnt at the bottom.
7. Cover the pot with lid. Remove lid when water boils again. Pour in a bowl of water. Cover with the lid again. Remove lid when water boils again. Cook over medium heat for 2min and it is done.

Chives and Pork Dumplings
韭菜豬肉水餃 (P20)
Time: 8-10 mins   Serve: 4-6

Ingredients

Cold dough
500gm Plain flour, 250ml Water

Stuffing
400gm Pork filling (please refer to page 12), 500gm Chives, chopped

Procedures
1. For making of cold dough, please refer to page 17.
2. Combine all filling and mix well.
3. Remove a portion of dough from mixing bowl. Roll it into a cylinder as thick as a thumb. Divide into evenly sized pieces, about 19-22gm per piece. Slightly stretch out each dough and using rolling pin, roll flat and make edges thinner.
4. Put the wrapper on table, spoon filling onto the middle. Press a bit.
5. Fold the wrapper in half. Crimp the edges and pinch together to seal.
6. Bring water to a boil. Stir water to form swirls. Place dumplings inside and stir slightly a bit to avoid them getting burnt at the bottom.
7. Cover the pot with lid. Remove lid when water boils again. Pour in a bowl of water. Cover with the lid again. Remove lid when water boils again. Cook over medium heat for 2min and it is done.

Mutton Dumplings
羊肉水餃 (P22)
Time: 8-10 mins   Serve: 4-6

Ingredients
Cold dough, 500gm Plain flour, 250ml Water

Stuffing
600gm Mutton Filling (please refer to page 13 for detail), 2 tablespoon Sesame oil, added lastly

Procedures
1. Combine all ingredients for Filling. Add sesame oil and mix well. Set aside.
2. For making cold dough, please refer to page 17. Ferment for 15min.
3. Remove a portion of dough from mixing bowl. Roll it into a cylinder as thick as a thumb. Divide into evenly sized pieces, about 19-22gm per piece. Slightly stretch out each dough and using rolling pin, roll flat and make edges thinner.
4. Put the wrapper on table, spoon filling onto the middle. Press a bit.
5. Fold the wrapper in half. Crimp the edges and pinch together to seal. Bring water to a boil. Stir water to form swirls. Place dumplings inside and stir slightly a bit to avoid them getting burnt at the bottom.
6. Cover the pot with lid. Remove lid when water boils again. Pour in a bowl of water. Cover with the lid again. Remove lid when water boils again. Cook over medium heat for 2min and it is done.

Three Treasures Dumplings
三鮮盒子 (P24)
Time: 8-10 mins   Serve: 4-6

Ingredients

Cold dough
500gm Plain flour, 250ml Water

Stuffing
300gm Pork front leg, minced, 200gm Fresh shrimp meat, drained to dry, 50gm Soaked fungus, chopped and to be added lastly, 200gm Chives, chopped, 50ml Oil, to be combined with chives, 2 Eggs, stir-fried, chopped and to be added lastly

Seasonings
3 tablespoon Oil, 1 tablespoon Granulated sugar, 1 tablespoon Abalone sauce, 150ml Peanut oil, 100ml Cold water, 2 tablespoon Sesame oil

Procedures
1. For making of cold dough, please refer to page 17. Then ferment for 15min.
2. Add oil, sugar, peanut oil, abalone sauce and 50ml of water in minced pork. Stir and throw it against side of bowl until combined sticky and compacted. Add remaining 50ml water, repeatly stir and throw until sticky and compacted. Sprinkle in chopped chives, eggs and sesame oil. Combine and mix well. Add remaining ingredients. Mix altogether and refrigerate.
3. Remove a portion of dough from mixing bowl. Roll it into a cylinder as thick as a thumb. Divide into evenly sized pieces, about 19-22gm per piece. Slightly stretch out each dough and using rolling pin, roll flat and make edges thinner.
4. Put a wrapper on table, spoon filling onto the middle. Press a bit. Top with other wrapper. Crimp the edges and pinch together decoratively with fingers to seal, like a box.
5. Bring water to a boil. Stir water to form swirls. Place dumplings inside and stir slightly a bit to avoid them getting burnt at the bottom.
6. Cover the pot with lid. Remove lid when water boils again. Pour in a bowl of water. Cover with the lid again. Remove lid when water boils again. Cook over medium heat for 2min and it is done.

Shanghai White Cabbage with 3 Treasures Dumplings
津白三鮮水餃 (P26)
Time: 8-10 mins   Serve: 4-6

Ingredients

Cold dough
500gm Plain flour, 250ml Water

Stuffing
300gm Pork Filling (please refer to page 12), minced, 300gm Fresh shrimps meat, drained to dry, 200gm Shanghai white cabbage, 2 Eggs, stir-fried and chopped, 100gm Soaked fungus, shredded

Seasonings
1/4 teaspoon fine salt, 1 tablespoon Sesame oil, 1 tablespoon Abalone sauce

Procedures

1. For making cold dough, please refer to page 17. Then ferment for 10-20min.
2. Combine all ingredients. Add in all seasonings and refrigerate.
3. Remove a portion of dough from mixing bowl. Roll it into a cylinder as thick as a thumb. Divide into evenly sized pieces, about 19-22gm per piece. Slightly stretch out each dough and using rolling pin, roll flat and make edges thinner.
4. Put a wrapper on table, spoon filling onto the middle. Press a bit.
5. Fold the wrapper in half. Crimp the edges and pinch together to seal.
6. Bring water to a boil. Stir water to form swirls. Place dumplings inside and stir slightly a bit to avoid them getting burnt at the bottom.
7. Cover the pot with lid. Remove lid when water boils again. Pour in a bowl of water. Cover with the lid again. Remove lid when water boils again. Cook over medium heat for 2min and it is done.

Chives and Fish Dumplings
韭菜魚肉水餃 (P28)
Time: 8-10 mins   Serve: 4-6

Ingredients

Cold dough
500gm Plain flour, 250ml Water

Stuffing
300gm Fish Filling (please refer to page 14), 300gm Chives, shredded, 2 tablespoon Ginger and green onion oil

Procedures

1. Refer to page 17 for procedures of making cold dough. Then ferment for 10-20 min.
2. Combine well all ingredients and mix them well. Refrigerate.
3. Remove a portion of dough from mixing bowl. Roll it into a cylinder as thick as a thumb. Divide into evenly sized pieces, about 19-22gm per piece. Slightly stretch out each dough and using rolling pin, roll flat and make edges thinner.
4. Put a wrapper on table, spoon filling onto the middle. Fold the wrapper in half. Press edges together to seal. Moisten one end and fold so that 2 ends join and overlap, like a ring.
5. Bring water to a boil. Stir water to form swirls. Place dumplings inside and stir slightly a bit to avoid them getting burnt at the bottom.
6. Cover the pot with lid. Remove lid when water boils again. Pour in a bowl of water. Cover with the lid again. Remove lid when water boils again. Cook over medium heat for 2min and it is done.

Vegetable Potstickers
花素鍋貼 (P30)
Time: 8-10 mins   Serve: 4-6

Ingredients

Combined dough
300gm Hot dough (please refer to page 17), 300gm cold dough (please refer to page 17)

Stuffing
1200gm Shanghai pak choi, 100gm Soaked fungus, chopped, 100gm Scallions, 4 Eggs, stir-fried and chopped

Seasonings
1 teaspoon Fine salt, 50gm Jett brand lard, 80ml Oil, 1 teaspoon Chicken powder, 1 tablespoon Abalone sauce, 1 tablespoon Sesame oil, to be added lastly

Procedures

1. Combine all ingredients. Add seasonings except sesame oil and mix well. Then add sesame oil. Mix well. Refrigerate.
2. Combine hot dough and cold dough. Rub and knead. Add plain flour during kneading process, to get slight sheen surface.
3. Remove a portion of dough from mixing bowl. Roll it into a cylinder as thick as a thumb. Divide into evenly sized pieces, about 19-22gm per piece. Slightly stretch out each dough and using rolling pin, roll flat and make edges thinner.
4. Put a wrapper on table, spoon filling onto the middle. Press a bit.
5. Fold the wrapper in half. Press edges together. Crimp the edges and pinch together to seal. Place dumplings in the same direction inside a skillet. Pan-fry over high heat for 1min. Pour in 300ml water. Cover with lid.
6. When water almost fully evaporates, add 2 tablespoon of oil. Cover the lid. Turn to medium heat. Hold the skillet and shake continuously for each dumpling to be cooked in uniform amount of heat. When water fully evaporates, gently pry up bottom of dumplings. If they turn golden, remove from skillet and ready to serve.

Mutton and Zucchini Potsticker
羊肉翠瓜鍋貼 (P32)
Time: 8-10 mins   Serve: 4-6

Ingredients

Combined dough
300gm Hot dough (please refer to page 17), 400gm Cold dough (please refer to page 17)

Stuffing
400gm Mutton filling (please refer to page 13), 1.5 catties (900gm) Zucchini, shredded, 1 teaspoon Fine salt

Procedures

1. Sprinkle fine salt to shredded zucchini and mix well. Marinate for 3min to release juice. Chop a little. Combine well with mutton Filling and refrigerate.
2. Combine hot dough and cold dough. Rub and knead. Add plain flour during kneading process, to get slight sheen surface. Remove a portion of dough from mixing bowl. Roll it into a cylinder as thick as a thumb. Divide into evenly sized pieces, about 19-22gm per piece. Slightly stretch out each dough and using rolling pin, roll flat and make edges thinner. Place a spoon of Filling on top.
3. Fold the wrapper in half. Crimp the edges and pinch together decoratively with fingers to seal.
4. Heat skillet and add oil. Place dumplings in the

same direction inside a skillet. Cover with lid. Pan-fry over high heat for 1min.

5. When water almost fully evaporates, add 2 tablespoon of oil. Cover the lid. Turn to medium heat. Hold the skillet and shake continuously for each dumpling to be cooked in uniform amount of heat. Keep frying until bottoms turn golden. Remove from skillet and ready to serve.

Traditional Potstickers
傳統鍋貼 (P34)
Time: 10 mins   Serve: 4-6

Ingredients

Combined dough

100gm Hot dough (please refer to page 17), 500gm Cold dough (please refer to page 17)

Stuffing

600gm Beef Filling (please refer to page 12), 200gm Onion, chopped

Thickening

300ml Water, 1/2 tablespoon Plain flour, 1/3 teaspoon salt

Procedures

1. Combine beef Filling with chopped onion. Refrigerate.

2. Combine hot dough and cold dough. Rub and knead. Add plain flour during kneading process, to get slight sheen surface.

3. Remove 25gm of dough from mixing bowl. Slightly stretch out each dough and using rolling pin, roll flat and make edges thinner. Spoon Filling of 20gm on top.

4. Fold the wrapper in half. Moisten one end and fold so that 2 ends join and overlap.

5. Combine ingredients for thickening and set aside.

6. Heat skillet and add oil. Place dumplings in the same direction inside a skillet. Pan-fry over high heat for 1min. Pour in 300ml of thickening. Cover the lid. When water almost fully evaporates, turn to medium heat. Keep shaking the skillet to allow each dumpling getting uniform amount of heat. After 1min, pour in 1 tablespoon from the side of skillet.

7. Keep shaking the skillet for each dumpling to move. Put a plate with bottom facing up to cover dumplings. Until they are golden, hold the plate and skillet. Flip them over for dumplings setting on the plate and ready to serve.

Home-make Pan-fried Breads and Pan-fried Dumplings
家常餅與餡餅類

Tianjin Pan-fried Bread
天津水麵餅 (P36)
Time: 3-5 mins   Made: 15

Ingredients

600gm Plain flour, 350ml Water, 1 tablespoon Peanut oil

Procedures

1. Make a well in the center, add some portion of water, mix with flour, knead and rub. Repeat

procedures until all water utilized and slight sheen surface. Cover with plastic cling for fermentation of 20min. Rub and knead again until slight sheen surface. Cover dough again for 10min of fermentation. Set aside.

2. Sprinkle flour on top of working surface. Put fermented dough on top. Use rolling pin, roll flat to thin round.

3. Push peanut oil on top. Roll from outside to inside. Cut into even pieces as per preferred portions.

4. Use a rolling pin to roll flat to about 1cm flat.

5. Heat the skillet with oil. Pan-fry until golden.

Stewed bread with Shanghai White Cabbage and Shredded Pork
津白肉絲燜餅 (P38)
Time: 10 mins   Serve: 4-6

Ingredients

Cake

600gm Plain flour, 350ml Water, 1 tablespoon Peanut oil

Stuffing

300gm Shredded pork, Half stalk Tianjin pak choi, 6 Soaked shiitakes, 100gm Minced scallions, 400ml Broth

Seasonings

Enough salt, Enough MSG, Enough dark soy sauce

Procedures

Bread

Refer to "Tianjin Pan-fried Bread"

Stewing

1. Divide home-made pan-fried cake into 3 portions and shred. Set aside.

2. Rinse pak choi and shred coarsely. Shred shiitakes and mince scallions.

3. Heat the wok with oil. Stir fry scallions a little. Add shredded pork to fry until fragrant. Add shiitake and pak choi. Fry over high heat until soften.

4. Add seasonings. Bring it to a boil and quick pour in cakes. Keep frying until sauce fully being absorbed.

Pan-fry Scallion Bread
葱花餅 (P40)
Time: 8-10 mins   Made: 16-18

Ingredients

Dough

600gm Plain four, 1 tablespoon Oil, 350ml Water

Stuffing

Enough Caul fat or pork fat, 2-3 tablespoon Scallions, 1/2 teaspoon Salt, 1/4 teaspoon Five-spice powder

Procedures

1. Mix flour with oil. Pour in water slowly and in-corporate until loose, shaggy mass. Rub and knead until slight sheen surface. Cover with plastic wrap or wet towel for fermentation of 20min. Rub and knead again until slight sheen surface. Cover with plastic wrap or wet towel and ferment for 10min.

2. Transfer the dough to a lightly floured working table. Use rolling pin and roll the dough into a thin.

3. Brush top with a film of peanut oil, place caul fat or pork fat on top. Sprinkle evenly with minced scallions, salt and 5-spice powder. Roll up dough from the outer end into a cylinder. Cut into even pieces in desired size.

4. Work with one piece at a time. Seal openings of both ends. Press slightly to flatten. Roll into a 1-cm-thick thin.

5. Heat the skillet over low heat. Add 50ml of peanut oil. Add first bread to the pan, cook until both sides are golden. Remove to paper towel-lined plate.

Pan Cakes (for serving with Peking duck)
單餅或北京填鴨餅 (P42)

Time: 2-3 mins   Made: 20

Ingredients

Combined dough
400gm Hot dough (please refer to page 17), 100gm Cold dough (please refer to page 17), Enough flour (for in-corporating dough)

Procedures

1. In between the cold dough and hot dough, sprinkle flour. Combine well. Rub and knead until slight sheen surface. Cover with plastic wrap or wet towel. Ferment for 15min. Dough is done.

2. Dust working surface with flour. Place the dough on top. Roll into a cylinder. Divide evenly into small pieces, each about 20gm. Roll flat to thin round.

3. Heat the skillet with a bit of oil over medium heat. Pan-fry pan cakes until golden on both sides and bubbles form.

Semi-deep Fried Stuffed Scallion Breads
葱油餅 (P44)

Time: 10 mins   Made: 30-35

Ingredients

Dough
600gm Plain flour, 2 tablespoon Oil, 360ml Water

Stuffing
600gm Minced scallions, 200gm Jatt brand lard, 25gm Sesame oil, 1 teaspoon Fine salt, 1 tablespoon MSG

Procedures

1. Mix sesame oil and lard. Sprinkle in minced scallions. Add fine salt and MSG, mix well. Salt will release juice from scallions, sieve.

2. Mix flour with oil. Pour in water slowly and in-corporate. Rub and knead until slight sheen surface. Cover with plastic wrap or wet towel for fermentation of 20min. Use rolling pin and roll the dough into a square thin. Sprinkle flour on top. Roll up dough from the far end into a cylinder.

3. Cut into even pieces, each about 50gm. Press slightly to flatten. Roll into round thin. Spoon 30gm of Filling onto the middle. Roll up from inner side into a cylinder. Use your hands to hole each end. Moisten one end and fold so that 2 ends join and overlap, like a ring. Overlapping part faces down.

4. Heat the skillet with enough oil to cover 2/3 of rings until 70% boiled. Place bread inside the skillet. Cook until golden on both sides. Serve hot.

Great Scallion Bread (Fu Bread)
葱花大餅 / 福字餅 (P46)

Time: 10 mins   Made: 6-8

Ingredients
Enough Sesame, for garnish

Dough
600gm Plain flour, 1 tablespoon Granulated sugar, 4gm (1/2 tablespoon) Yeast, 350gm Water

Stuffing
1 teaspoon Sichuan pepper salt, 100gm White sesame, 4 tablespoon Peanut oil, 200gm Minced scallions, 50ml Egg pulp

Procedures

1. Mix flour with yeast and peanut oil. Add water several times. Rub and knead with slight sheen surface. Cover with plastic wrap or wet towel to ferment for 10min.

2. Dust working surface with flour. Put the dough on top. Rub and knead again until a slight sheen surface.

3. Use rolling pin and roll the dough into a square thin. Brush surface with oil. Sprinkle Sichuan pepper salt, and then dust with flour. Sprinkle minced scallions. Roll up dough from the far end toward inner end into a cylinder.

4. Cut into even pieces in desired size. Fold both ends down to meet the bottom. Roll into round thin. Brush with egg pulp. Sprinkle some sesame seeds all over.

5. Heat wok with oil. Pan-fry until both sides get slight golden. Pour in 100ml of water. Bring it to a boil, add oil and pan-fry until golden.

Golden Great Bread
黃金大餅 (P48)

Time: 15 mins   Made: 8-10

Ingredients
Enough egg pulp, for brushing the tops

Dough
600gm Plain flour, 1 teaspoon Yeast, 1 teaspoon Baking powder, 1/2 teaspoon Fine salt, 1 tablespoon Granulated sugar, 2 tablespoon Oil, 300gm Water, to be added lastly

Stuffing
400gm Shredded turnips, 2 teaspoon Fine Salt, for marinating turnips, 2 tablespoon Oil, for stir frying turnips, 200gm Minced green onions, 50gm Minced ham, 100gm white sesame

Seasonings
7-8gm Salt, 1 tablespoon Chicken powder, 2 tablespoon Granulated sugar, 200gm Water, Enough cornstarch

Procedures

1. Mix shredded turnips with salt for 5min to release juice. Squeeze and drain.

2. Heat the wok and add oil. Stir fry for 3min. Sprinkle in salt and sugar. Stir well. Mix water with cornstarch to thicken. Sprinkle minced green onions and ham. Remove from heat. Let cool and refrigerate.

3. Mix ingredients of dough together well except water. Slowly pour water. In-corporate until dough is formed. Cover with wet towel. Ferment for 1hr. Rub and knead again until a slight sheen surface.

4. Divide evenly into small pieces, each about 100-150gm. Roll flat into thins. Spoon 60-80gm of Filling onto the middle. Moisten edge with water. Cover with other thin op top. Press edge.

5. Use the edge of a bowl to cut unnecessary dough around edges. Brush surface with egg pulp. Sprinkle sesame all over. Steam over high heat for 10min. Heat the wok with oil, deep fry until golden.

Danish Swirls in Chinese Style
圈圈餅（草帽餅）(P50)

Time: 10 mins   Made: 4

Ingredients

Dough
500gm Cold dough (please refer to page 17), 150gm Hot dough (please refer to page 17), 2 tablespoon Oil

Suffings
Enough Sesame oil, Enough Sichuan pepper salt, for dusting, Enough flour, for dusting

Procedures

1. Combine cold dough with hot dough and oil. Rub and knead until silky and soft dough.

2. Divide into 4 even portions. Roll one portion into rectangular thin. Brush surface with sesame oil. Sprinkle Sichuan pepper salt. Then dust on top with flour.

3. Fold the thin into a hand-held fan. Stretch. Hold each end by one hand. Lock left hand. Right hand revolves until swirl forms.

4. Heat wok with oil. Pan-fry until both sides golden. Remove from heat. Place on cutting board. Cover with wet towel. Slightly flatten with Chinese knife.

Clay Oven Breads
叉子燒餅 (P52)

Time: 10 mins   Made: 30

Ingredients

Enough egg pulp, for brushing the tops, Enough sesame, for garnish

Dough
600gm Plain flour, 2 tablespoon Oil, 350ml Water

Puff Pastry
80gm Plain flour, 50ml Oil

Procedures

1. Mix flour, oil and water. In-corporate until dough formed. Ferment for 30min and set aside. Combine ingredients of puff pastry and mix until paste.

2. Flatten fermented dough to rectangular thin by using rolling pin. Top with puff pastry paste. Roll up dough from the far end into a cylinder. Divide evenly into small pieces by knife.

3. Take one piece. From 1/3 of position, use rolling pin to roll a bit to form a slope. Continue to roll. From where the cutting edge is, roll flat 1/4 into a slope. Fold. Change position and roll flat.

4. After all small pieces are done, flatten by hand. In the middle part, roll flat once. Change position and roll flat at the front and back. Brush tops with egg pulp. Sprinkle sesame allover. Pre-heat oven to 240°C and bake for 10min.

Chinese Pita Bread soaked in Lamb Soup
羊肉湯泡   (P54)

Time: 30 mins   Serve: 4-6

Ingredients

400gm Fermented dough (please refer to page 15), 200gm Plain flour

Lamp Soup
900gm Leg of mutton (mutton slices), Half stalk Tianjin white cabbage, shredded, Enough Sichuan pepper, Enough Scallions, 800ml Water, 5 stalks Coriander

Procedures

1. Combine fermented dough with flour, rub and knead until fine dough formed. Divide into 5 equal portions. Flatten.

2. Put dough in a pan, fry until both sides get golden. Pre-heat oven to 120°C and bake for 15min.

3. Stir fry Sichuan pepper and scallions a little. Add mutton until fragrant. Pour in water and bring it to a boil.

4. Add Tianjin white cabbage, cook for 10min.

5. Rinse coriander thoroughly. Chop into sections. Sprinkle into the mutton soup.

6. When serving the soup, dip in small pieces of pita bread.

Pumpkin and Pork Stuffed Pan-fried Breads
南瓜豬肉餡餅 (P56)

Time: 6-8 mins   Made: 18

Ingredients

Dough
600gm Plain flour, 1 teaspoon Yeast, 1 tablespoon Granulated sugar, 1 tablespoon Oil, 1 Egg, 350ml Water

Stuffing
300gm Pork Filling (please refer to page 12), 600gm Pumpkin

Procedures

1. Skin and core pumpkin. Rinse. Shred and then apply a few cuts. Mix well with pork Filling.

2. Combine flour with yeast, sugar, oil and egg. Mix well. Add water several times and continue to rub and knead until slight sheen surface. Cover with wet towel. Ferment in room temperature for 10min.

3. Divide into several pieces, each about 50gm. Roll flat to round thins. Place a spoonful of 40gm filling in the centre. Shape to form buns. Opening faces down. Flatten.

4. Heat wok with oil. Pan-fry until golden. After 2-3min, turn it over and pan-fry other side until golden.

Stuffed Celery and Pork Pan-fried Breads
芹菜豬肉餡餅 (P58)

Time: 8-10 mins   Made: 10

Ingredients

Dough
400gm Plain flour, 1 tablespoon Oil, 220ml Water

Shortening
200gm Plain flour, 100gm Oil

Stuffing

300gm Pork Filling (please refer to page 12), 400gm Celery

Procedures

1. Chop celery and mix with pork Filling. Set aside.

2. Combine ingredients of dough. Rub and knead until soft and smooth dough. Cover with wet towel for fermentation of 15min.

3. Mix ingredients for shortening. Mix to form paste. Set aside.

4. Place dough on a working surface. Rub and knead until sheen surface. Use rolling pin and roll the dough into a square thin. Put shortening on top. Roll up dough from the far end into a cylinder. Cut into small pieces, each about 50gm. Flatten. Place a spoonful of 40gm fillings in the centre. Begin to form a ball by pulling sides up and over the filling. Pinch and pleat the edges and make sure ball is sealed. Flatten.

5. Heat wok with oil. Pan-fry until both sides golden.

Stuffed Zucchini and Shrimp Pan-fried Beards

翠瓜 · 鮮蝦仁 · 雞蛋餡餅 (P60)

Time: 30 mins   Serve: 4-6

Ingredients

Dough

600gm Plain flour, 1 tablespoon Oil, 350ml Water

Stuffing

3 Zucchini, 1 stalk Scallion, 200gm Fresh shrimps, 1 teaspoon Salt, 4 Eggs, fried and chopped

Seasonings

1 teaspoon Fine salt, 2 tablespoon Abalone sauce, 4 tablespoon Oil, 1 tablespoon Sesame oil

Procedures

1. Rinse zucchini, shred by using shredder. Sprinkle salt and marinate for 3 min to release juice. Set aside.

2. Skin scallion. Rinse and chop. Wash and rinse shrimps thoroughly. Chop if necessary.

3. Mix all ingredients. Add salt, abalone sauce and oil. Combine well. Sprinkle 1 tablespoon of sesame oil. Refrigerate.

4. Combine flour with oil. Add water slowly and mix in one direction until loose, shaggy mass. Rub and knead until it is relatively smooth and has a very slight sheen. Cover with wet towel. Ferment for 20 min.

5. Divide dough into smaller pieces, each about 50gm. Slightly stretch out each dough and using rolling pin, roll flat and make rounds. Spoon Filling of 40gm on top. Fold to half and seal opening.

6. Heat skillet over medium heat. Add dumplings and over the lid for 1 min. Turn them over. Cover the lid and cook until done.

Chives Dumplings

韭菜素盒子 (P62)

Time: 30 mins   Serve: 4

Ingredients

600gm Plain flour, 1 tablesppon Oil, 350ml Water

Stuffing

500gm Chives, chopped, 50ml Oil, 200gm Fresh shrimps, fried, 4 Eggs, fried, 2 Chinese fritters, chopped

Seasonings

1/2 teaspoon Salt, 1 tablespoon Chicken powder, 2 tablespoon Sesame oil

Procedures

1. Mix chives and oil. Then combine with rest of Filling ingredients and seasonings, mix well. Set aside.

2. Mix oil with flour. Divide 350ml of water into 3 portions. Pour in first portion of water and stir flour. Rub and knead. Repeat steps until all portions of water done. Rub and knead until silky surface. Cover with wet towel and ferment for 20min.

3. Remove dough from mixing bowl. Roll into a cylinder. Divide evenly into small pieces, each about 50gm. Roll flat to thin oval wrappers.

4. Place each wrapper in the centre of left hand. Put on top 40gm of filling. Use the right hand to fold wrapper into half. Use both hands to seal opening. Place dumpling on a slightly floured surface. Cut unnecessary edges by a bowl.

5. Heat the skillet over low heat. Pan-fry without oil until slight brown.

Buns and Steamed Rolls
包點和花卷

Tianjin Buns Left Alone

誰不理包子 (P64)

Time: 60 mins   Serve: 6

Ingredients

Dough

600gm Plain flour, 1 teaspoon Yeast, 2 teaspoon Baking powder, 1 tablespoon Granulated sugar, 2 tablespoon Oil, 350ml Water

Stuffing

500gm Pork front leg lean, 150gm Lean pork, 100gm Scallions

Seasonings

1 tablespoon Soy sauce, 50ml Sesame oil, 25gm Sweet bean paste, 1 teaspoon Fine salt, 1 tablespoon Granulated sugar, 1 tablespoon MSG, 300ml Ham broth, 150ml Peanut oil, 2 tablespoon Cooking wine

Procedures

Stuffing

1. Rinse pork front leg and mince. Rinse lean pork and chop into small pieces like peanuts. Marinate with soy sauce for 10 min. Fry until cooked. Set aside. Rinse scallions and chop, add sesame oil and combine well.

2. Add salt, sugar, MSG, cooking wine, peanut oil and some broth in minced pork front leg. Stir at one direction and throw it against the side of bowl until combined sticky and compacted.

3. Pour in rest of broth, repeat stirring and throwing until combined sticky and compacted. Add sweet bean paste, scallions and cooked diced pork. Mix well and refrigerate.

Wrapper

1. Mix flour with yeast, baking powder, sugar and oil. Add and merge some portion of water. Rub and knead. Repeat steps until all portions of water done. Rub and knead until silky surface. Cover with wet towel and ferment for 1 hour.
2. Remove wet towel, rub and knead until silky surface. Cover again with wet towel for 10min. Roll into a cylinder. Divide evenly into small pieces, each about 35gm. Slightly stretch out each dough and using rolling pin, roll them flat to rounds and make edges thinner.
3. Place Filling of 35gm to each round. Seal opening. Set aside for 15min.
4. Pour water in a steamer. Put wet cloth or parchment paper underneath buns. Steam for 15min.

Provenance

Stuffing to be used for buns and soup dumplings are different. Fillings as such Meat, melting gold and bean paste can be used for buns while they cannot apply to soup dumplings. There are great varieties of fillings for buns, especially buns available in Tianjin. One of them, a very famous one, is called "Buns that dogs ignore" or "Buns left alone". Once upon a time, an owner of a restaurant sat outside of his restaurant as there was no customer. His friend encountered him and asked if he had good business. The restaurant owner showed a funny face and asked him to go inside. The friend joked by saying the buns that the restaurant produced were not good, wrappers too thick and fillings too little, even dogs ignore. The restaurant owner liked the name "dogs ignore" so much. He thus improved the quality of the buns and named the buns as "Buns that dogs ignore". From then on, the buns became so popular.

Shanghai Pan-fried Buns
生煎包 (P66)
Time: 60 mins   Serve: 6

Ingredients

250ml Water, for pan-frying buns, 1 tablespoon Sesame oil, to be added lastly

Dough
600gm Plain flour, 1/2 teaspoon Yeast, 1 tablespoon Granulated sugar, 2 tablespoon Soy sauce, 50gm Corn starch, 330ml Water, to be added lastly

Stuffing
600gm Pork front leg, minced, 50gm Scallions, chopped, 200gm Pork gelatin, pinched and added lastly (please refer to page 73)

Seasonings
1 teaspoon Fine salt, 1 tablespoon Granulated sugar, 1 tablespoon MSG, 100ml Ham broth, 150gm Peanut oil, 3 tablespoon Cooking wine

Procedures

1. Mix Filling ingredients with seasonings. Stir and pound in one direction until sticky and elastic. Merge with pork gelatin and mix well. Sprinkle sesame oil. Refrigerate.
2. Mix ingredients for dough. Add and merge some portion of water. Rub and knead. Repeat steps until all portions of water 330ml done. Rub and knead until silky surface. Cover with wet towel and ferment for 1 hour.
3. Remove wet towel, rub and knead until silky surface. Cover again with wet towel for 15min. Roll into a cylinder. Divide evenly into small pieces, each about 35gm. Slightly stretch out each dough and using rolling pin, roll them flat to rounds and make edges thinner.
4. Place Filling of 35gm to each round. Crimp the edges and pinch together to seal.
5. Heat the skillet and then add oil. When oil fully heated, put in buns. Pour in 250ml of water. Cover with lid and cook until all water evaporates. Sprinkle sesame oil and continue to cook until bottom of buns golden.

Vegetable Buns
素菜包 (P68)
Time: 40 mins   Serve: 4

Ingredients

Dough
600gm Plain flour, 1 teaspoon Yeast, 2 teaspoon Baking powder, 1 tablespoon Granulated sugar, 2 tablespoon Oil, 350ml Water

Stuffing
30gm Soak shiitake, minced, 1 Stalk scallion, 1 slice Ginger, 4 catties (2400gm) Shanghai pak choi

Seasonings
1 teaspoon (7-8gm) Fine salt, 1 tablespoon Granulated sugar, 1 tablespoon Chicken powder, 1 tablespoon Sesame oil, 50ml Oil

Procedures

1. Pan-fry scallion and ginger with oil until fragrant. Add shiitake and continue to cook until water evaporates. Remove to the plate and separate shiitake from ginger and scallion.
2. Rinse Shanghai pak choi. Blanch and then rinse with cold water. Chop and squeeze out juice with a towel until weights 500-600gm. Add seasonings. Mix with shiitake and refrigerate.
3. Mix flour, yeast, baking powder, sugar and oil. Merge some portion of water. Rub and knead. Repeat steps until all water done. Rub and knead until silky surface. Cover with wet towel and ferment for 1 hour.
4. Remove wet towel, rub and knead until silky surface. Cover again with wet towel for 10min. Roll into a cylinder. Divide evenly into small pieces, each about 35gm. Slightly stretch out each dough and using rolling pin, roll them flat to rounds and make edges thinner.
5. Place Filling of 35gm to each round. Crimp the edges and pinch together to seal. Set aside for 15min.
6. Pour water in a steamer. Put wet cloth or parchment paper underneath buns. Steam 10min over high heat.

Shandong Giant Buns
山東大包 (P70)
Time: 50 mins   Serve: 4-6

Ingredients

Dough
900gm Plain flour, 2 teaspoon Yeast, 2 teaspoon Bakign

powder, 1 tablespoon Granulated sugar, 2 tablespoon Oil, 500ml Water

Stuffing

150gm Lean pork front leg, diced and pan-fried, 800gm Tianjin white cabbage, 1 stalk Scallion, chopped, 20gm Coriander, cut into 1-cm-section, 150gm Soaked fungus, chopped, 150gm Soaked cellophane noodles, cut into sections, 2 tablespoon Sweet bean paste

Seasonings

2 teaspoon Salt, 1/2 teaspoon Fine salt, 2 teaspoon Granulated sugar, 1 tablespoon Chicken powder, 50ml Peanut oil, 1 tablespoon Sesame Oil, 1/4 teaspoon 5-spice powder

Procedures

1. Rinse Tianjin white cabbages and chop. Marinate with 2 teaspoon of salt. Squeeze out juice with towel.
2. Mix Filling ingredients with seasonings and combine well. Refrigerate.
3. Combine flour with yeast, baking powder, sugar and oil. Merge some portion of water. Rub and knead. Repeat steps until all water done. Rub and knead until silky surface. Cover with wet towel and ferment for 10 min.
4. Roll into a cylinder. Divide evenly into small pieces, each about 50gm. Slightly stretch out each dough and using rolling pin, roll them flat to rounds and make edges thinner. Place Filling of 60gm to each round. Crimp the edges and pinch together to seal. Set aside for 15min.
5. Pour water in a steamer. Put wet cloth or parchment paper underneath buns. Steam 15min over high heat.

Xiaolongbao
小籠包 (P72)
Time: 60 mins   Serve: 6

Ingredients

Dough

600gm "Pak Choi" Brand flour, 2 tablespoon Oil, 300ml Water

Pork gelatin

600gm Pig skin, 4-5 stalks Scallion, 1 slice Ginger, 100ml Cooking wine, 2700ml Water

Stuffing

600gm Lean pork front leg, minced, 100gm Scallion, chopped, 400gm Pork gelatin, pinched and added lastly, 1 tablespoon Sesame oil, added lastly

Seasonings

1 teaspoon Fine salt, 2 tablespoon Granulated sugar, 1 tablespoon MSG, 100ml Peanut oil, 2 tablespoon Cooking wine, 1/2 tablespoon Dark soy sauce

Procedures

1. Rinse and wash pig skin thoroughly. Blanch and then mince. Add the rest of ingredients for pork gelatin. Bring to a boil over high heat for 5 min. turn the heat down to low and boil for 2.5 hr. Sieve and let cool. Refrigerate 20 hr. Pork gelatin will form.
2. Mix minced pork with seasonings. Repeatly stir and throw until sticky and compacted. Merge with minced pork gelatin. Sprinkle sesame oil and

refrigerate.
3. Combine flour with oil. Pour 300ml of water. Rub and knead until dough forms. Cover with wet towel and ferment for 20min. Remove towel, rub and knead until silky surface.
4. Roll into a cylinder. Divide evenly into small pieces, each about 16gm. Slightly stretch out each dough and using rolling pin, roll them flat to rounds and make edges thinner. Place Filling of 20gm to each round. Crimp the edges and pinch together to seal.
5. Pour water in a steamer. Put wet cloth or parchment paper underneath buns. Bring water to a boil and steam 10min over high heat.

Longevity Peach Buns
壽桃百子壽桃 (P74)
Time: 10 mins   Serve: 10

Ingredients

Enough red food colouring, for garnish

Dough for plate

200gm Plain flour, 1/2 teaspoon Yeast, 1 teaspoon Baking powder, 1 tablespoon Granulated sugar, 1 tablespoon Oil, 100ml Water

Mini peaches

500gm Plain flour, 1 teaspoon Yeast, 2 teaspoon Baking powder, 1 tablespoon Granulated sugar, 1 tablespoon Oil, 250ml Water

Numerous generations peaches

300gm Plain flour, 1/2 teaspoon Yeast, 1 teaspoon Baking powder, 1 tablespoon Granulated sugar, 1 tablespoon Oil, 130ml Water

Stuffing for numerous generations peaches

30 units Bean paste, 30 units Chocolate

Procedures

Dough for plate

Mix flour with yeast, baking powder, sugar and oil. Merge with 100ml of water. Rub and knead until silky surface. Use a rolling pin to roll flat to a round sheet in 9 inches diameter. Steam until cooked.

Mini peaches:

1. Combine flour with yeast, baking powder, sugar and oil. Merge in 250ml of water. Rub and knead until silky surface. Divide into 8 equal portions, each about 25-30gm.
2. Use a bowl as the mould, wrap with plastic cling. Take 2/3 out of the remaining dough, shape into a peach. Put leaves on. Steam until cooked.

Numerous generations peaches

Numerous generations peaches: Mix flour, yeast, baking powder, sugar and oil. Merge with water, rub and knead until silky surface. Divide into 60 equal pieces, each about 10gm. Each wraps bean paste or chocolate until done. Shape into peaches. Steam until cooked. Set aside to cool down.

Assembly

Brush tops of numerous generations peaches with food colouring. Put them on the plate and fix position. Put mini peaches around.

Chinese Steamed Buns
刀切饅頭 (P76)

Time: 15 mins　　　　Serve: 6

Ingredients

600gm Plain flour, 1 teaspoon Yeast, 2 teaspoon Baking powder, 1 tablespoon Granulated sugar, 1 tablespoon Oil, 300ml Water

Procedures

1. Combine flour with yeast, baking powder, sugar and oil. Divide 300ml of water into 3 portions. Stir flour while pouring in first portion of water. Repeat steps until all portions of water done. Rub and knead until silky surface. Cover with plastic cling to ferment 30min.
2. Rub and knead the dough again until silky surface. Ferment 10min more.
3. Dust the working surface with flour. Use rolling pin and roll the dough into a square thin. Sprinkle flour on top. Roll up dough from the far end into a cylinder and use your hands to tighten. Press the end point.
4. Cut into pieces of desired size. Ferment 20min.
5. Pour water in a steamer. Put wet cloth or parchment paper underneath buns. Bring water to a boil and steam 15min over high heat.

Green Onion Rolls
葱花卷 (P78)

Time: 25 mins　　　　Serve: 6

Ingredients

Dough

600gm Plain flour, 1 teaspoon Yeast, 2 teaspoon Baking powder, 1 tablespoon Granulated sugar, 1 tablespoon Oil, 300ml Water

Stuffing

1/2 teaspoon Fine salt, 1 teaspoon pepper, 1/2 teaspoon 5-spice powder, 50gm Chopped green onions

Procedures

1. Mix flour, yeast, baking powder, sugar, oil and water. Rub and knead until dough forms with silky surface. Ferment 30min. Rub and knead again until silky surface, set aside.
2. Mix salt, pepper and 5-spice powder, set aside. Slightly dust the working area with flour. Divide dough into 2 equal chunks. Use rolling pin and roll the dough into a square thin. Brush on top with oil. In sequent, sprinkle mixture of above point 2, very thin layer of flour and chopped green onions.
3. Roll up dough from the far end into a cylinder and use your hands to tighten. Press the end point. Cut into sections. Use chopstick to press. Use both hands to stretch, pull and then twist. Both hands meet together. Seal both ends to form twisted roll.
4. Pour water in a steamer. Put wet cloth or parchment paper underneath buns. Bring water to a boil and steam 15min over high heat.

Hoof Rolls
牛蹄花卷 (P80)

Time: 40 mins　　　　Serve: 6

Ingredients

Dough

600gm Plain flour, 1 teaspoon Yeast, 2 teaspoon Baking powder, 1 tablespoon Granulated sugar, 1 tablespoon Oil, 300ml Water

Stuffing

1 teaspoon Sichuan pepper salt, 100gm Plain flour, Enough oil, for moistening dough

Procedures

1. Mix flour, yeast, baking powder, sugar and oil together. Stir flour while pouring in first portion of water. Repeat steps until all portions of water done. Rub and knead until silky surface. Ferment 30min.
2. Combine 100gm of flour with Sichuan pepper salt. Set aside.
3. Roll the dough into a cylinder. Divide into 18 pieces. Press each piece down and roll into round shape. Brush on top with oil and sprinkle a little mixture of flour and Sichuan pepper salt. Brush second thin layer of oil. Fold into halves. Cut 2/3 of length away from the ends, fold downward and pull the ends together. Seal well. Ferment 15-20min.
4. Pour water in a steamer. Put wet cloth or parchment paper underneath buns. Bring water to a boil and steam 10min over high heat.

Silver Thread Rolls
銀絲花卷 (P82)

Time: 40 mins　　　　Serve: 6

Ingredients

Strands of Dough

300gm Plain flour, 1/2 teaspoon Yeast, 3 tablespoon Granulated sugar, 1 tablespoon Oil, 130ml Water

Dough

300gm Plain flour, 1/2 teaspoon Yeast, 1 teaspoon Baking powder, 1 tablespoon Granulated sugar, 1 tablespoon Oil, 130ml Water

Procedures

Strands of Dough

1. Mix flour, yeast, sugar and oil. Stir flour while pouring in first portion of water. Repeat steps until all portions of water done. Rub and knead until silky surface.
2. Divide dough into 2 chunks. Place one chunk on the slightly floured working surface. Use rolling pin and roll the dough into a square thin. Dust with flour. Fold the thin into a hand-held fan and dust with flour on each folding.
3. Shred the hand-held fan. Keep moving the strands gently to avoid them sticking to each other. Brush strands with oil. Divide into 12 sections and refrigerate.

Wrapper

1. Combine flour with yeast, baking powder, sugar and oil. Merge in water, rub and knead until dough with silky surface.

2. Divide into 12 pieces equally. Slightly stretch out each dough and using rolling pin, roll them flat into square thins. Fold the side edges, put strands on top. Fold up any remaining edges. Press tightly and roll up into a cylinder. Ferment for 45 min

3. Pour water in a steamer. Put wet cloth or parchment paper underneath buns. Bring water to a boil and steam 10min over high heat.

Steamed Rolls
銀絲卷 (P84)

Time: 50 mins   Serve: 6

Ingredients

Strands of Dough
300gm Plain flour, 1/2 teaspoon Yeast, 3 tablespoon Granulated sugar, 1 tablespoon Oil, 130ml Water

Wrapper
300gm Plain flour, 1/2 teaspoon Yeast, 1 teaspoon Baking powder, 1 tablespoon Granulated sugar, 1 tablespoon Oil, 130ml Water

Procedures

Strands of Dough
1. Mix flour, yeast, sugar and oil. Stir flour while pouring in first portion of water. Repeat steps until all portions of water done. Rub and knead until silky surface.

2. Divide dough into 2 chunks. Place one chunk on the slightly floured working surface. Use rolling pin and roll the dough into a square thin. Dust with flour. Fold the thin into a hand-held fan and dust with flour on each folding.

3. Shred the hand-held fan. Keep moving the strands gently to avoid them sticking to each other. Brush strands with oil. Divide into 6 sections and refrigerate.

Wrapper
1. Combine flour with yeast, baking powder, sugar and oil. Merge in water, rub and knead until dough with silky surface. Divide into 6 pieces equally.

2. Slightly stretch out each dough and using rolling pin, roll them flat into square thins. Fold the side edges, put strands on top. Fold up any remaining edges. Press tightly and roll up into a cylinder. Ferment for 45 min.

3. Pour water in a steamer. Put wet cloth or parchment paper underneath buns. Bring water to a boil and steam 18min over high heat. Otherwise, deep-fry the bun until golden brown on surface.

Lotus Leaf Breads
荷葉餅 (P86)

Time: 20 mins   Serve: 4

Ingredients
200gm Fermented dough (please refer to page 15), Enough oil, Enough flour

Stuffing
Enough oil, 50gm Flour

Procedures
1. Roll dough into a cylinder. Divide into 10 pieces.

2. Slightly stretch out each dough and using rolling pin, roll them flat and make oval thins. Brush on top oil. Dust on top with flour. Fold 4 times.

3. Use knife to score oblique lines on top. Shape to lotus leaves. Ferment for 15min.

4. Bring water to a boil and steam 7min over high heat.

Steamed Rolls
豬膶花卷 (P88)

Time: 30 mins   Serve: 6

Ingredients
600gm Fermented dough (please refer to page 15), Enough oil, Enough flour

Stuffing
Enough oil, 50-100gm Flour

Procedures
1. Divide fermented dough into 3 equal portions

2. Use rolling pin and roll each portion into a 10 x 15 cm square thin. Brush on top oil and dust with flour. Fold 4-5 times.

3. Stretch each cylinder. Entwine 3 cylinders. Ends face down. Ferment 15min in room temperature.

4. Bring water to a boil and steam 10min over high heat.

Lotus Rolls
蓮花卷 (P90)

Time: 30 mins   Serve: 6

Ingredients
600gm Fermented dough (please refer to page 15), Enough oil, Enough flour

Stuffing
Enough oil, 50-100gm flour

Procedures
1. Divide fermented dough into 3 equal portions.

2. Roll out to form a thin round, about 20cm in diameter. Brush on top oil, and then sprinkle flour. Fold into halves. Brush oil again and fold into halves.

3. Cut into 3 portions. Stack them up. Use a chopstick to press down a line. Then press another line perpendicularly at 1/3 position and the third line at 2/3 position.

4. Seal the ends. Ferment 15min.

5. Bring water to a boil and steam 10min over high heat.

Miscellaneous
雜類

Steamed Vegetable Dumplings
花素蒸餃 (P92)

Time: 50 mins   Serve: 4

Ingredients

Wrapper
300gm Hot dough (please refer to page 17), 300gm Cold dough (please refer to page 17)

Stuffing

1200gm Dwarf pak choi, 1200gm Shanghai pak choi, 100gm Soaked fungus, chopped, 100gm Soaked shiitake, chopped, 100gm Soaked cellophane noodles, chopped

Seasonings

1 teaspoon Fine salt, 2 tablespoon Granulated sugar, 1 tablespoon MSG, 100ml Ginger scallion oil, 1 tablespoon Abalone sauce, 2 tablespoon Sesame oil

Procedures

1. Rinse and wash pak choi. Blanch. Put in cold water. Mince and then squeeze out juice by using a bag. Set aside.
2. Mix all Filling together. Add fine salt, granulated sugar, MSG, ginger scallion oil and abalone sauce. Mix well. Sprinkle sesame oil. Refrigerate.
3. Combine hold dough with cold dough. Rub and knead. Add flour, rub and knead again. Repeat procedures until dough turns slight sheen surface.
4. Divide into equal small pieces, each about 35gm. Slightly stretch out each dough and using rolling pin, roll them flat and make thin rounds. Spoon filling of 35gm onto the middle. Press the filling a bit. Pull sides up and over the filling. Press opening and seal.
5. Pour water in a steamer. Put wet cloth or parchment paper underneath buns. Bring water to a boil and steam 10min over high heat.

Huhhot Siumai
羊肉燒賣 (P94)

Time: 10 mins   Serve: 4

Ingredients

Wrapper

400gm Hot dough (please refer to page 17), 200gm Cold dough (please refer to page 17)

Stuffing

300gm Fresh mutton, 200gm Cooked glutinous rice, 20 Fresh shrimp

Seasonings

1 tablespoon Sesame oil, 1 tablespoon Soy sauce, 1/2 tablespoon MSG, 1/4 tablespoon Sichuan pepper powder

Procedures

1. Rinse mutton, diced. Add sesame oil to cooked glutinous rice when still hot. Peel shrimps but remain the tails on.
2. Add soy sauce and MSG to diced mutton and mix well. Sprinkle Sichuan pepper powder. Stir and throw it against side of bowl until combined sticky and compacted. Ad cooked glutinous rice. Mix well. Set aside at warm atmosphere.
3. Combine hot dough with cold dough, rub and knead dough turns slight sheen surface. Divide into small pieces, each about 20gm. Slightly stretch out each dough and using rolling pin, roll them flat and make thin rounds.
4. Use left hand to hold a thin round, spoon filling of 20gm and a shrimp onto middle. Gather up edges of wrapper, about 1/3 length of diameter and gentle pleat, to form a siumai shape.

5. Pour water in a steamer. Put wet cloth or parchment paper underneath buns. Bring water to a boil and steam 10min over high heat.

Zhajiangmian
炸醬麵 (P96)

Time: 50 mins   Serve: 4

Ingredients

Handmade Noodles

400gm Plain flour, 1/2 teaspoon Fine salt, 180ml Cold water, 350gm Potatoes, shredded and pan-fried, 350gm Green beans, pan-fried, 200gm Cucumber, shredded, 200gm Carrot, shredded, blanched, 4-5 stalks Chinese celery, 4 cloves of Garlic

Seasonings of mashed garlic

1/4 teaspoon of salt, 40ml Water, 1 tablespoon Vinegar, 1 teaspoon Sesame oil

Procedures

1. Press garlic and add salt to marinate for a while. Make mash. Add in water, vinegar and sesame oil, mix well.
2. Mix flour and fine salt. Merge in cold water. Rub and knead until dough formed. Cover with wet towel and ferment 30min.
3. Divide into 2 chunks. Rub and knead until slight sheen surface. Use sturdy rolling pin and roll the dough into a square thin. Dust with flour. Fold the thin into a hand-held fan and dust with flour on each folding. Cut into noodles. Move noodles gently to avoid them sticking to each other.
4. Bring water to a boil. Add noodles and use chopsticks to stir. Boil for 5-6min. Put noodles in 4 bowls separately. In sequential order, add mash garlic, shredded cucumbers, potatoes, carrot and green beans. Then it is done.

Sauce

500gm Lean pork, minced and then cooked, 250gm Sweet bean paste, 70gm Soaked fungus, 1 Egg, beaten

Seasonings (Sauce)

1 teaspoon salt, 1 tablespoon sugar, 50ml water or stock

Procedures

Heat wok with oil, add 500g minced pork, saute until change colour. Add sweet bean paste, stir fry until frangent, add seasoning and cook for 1 minute, stir in fungus shreds, then pour egg on top.

Super Halogen Noodles
大滷麵 (P98)

Time: 30 mins   Serve: 4

Ingredients

Noodles

600gm Plian flour, 1/2 teaspoon Fine salt, 250ml Cold water

Halogen gravy

150gm Lean pork, 1 tablespoon Soy sauce, for marinating pork, 200gm Prawn, sliced, 150gm Soaked and prepared Sea cucumbers, sliced, 80gm Soaked fungus, sliced, 50gm Soaked Daylily, trimmed, 3 Shiitakes, sliced, 250ml Water, 2 Eggs, added lastly, 1 teaspoon sesame oil, added lastly

Seasonings

1/2 tablespoon Dark soy sauce, 1 tablespoon Abalone sauce, 7-8gm Fine salt, 1 teaspoon MSG

Thickening

20gm Corn starch, 3 tablespoon Water

Procedures

1. Rinse and wash lean pork, shred and marinate with soy sauce. Set aside.
2. Add flour and fine salt together. Merge with cold water. Rub and knead until slight sheen surface. Cover with towel for fermentation of 30min.
3. Divide dough into 2 chunks. Use sturdy rolling pin and roll the dough into a square thin. Dust with flour. Fold the thin into a hand-held fan and dust with flour on each folding. Cut into noodles. Move noodles gently to avoid them sticking to each other.
4. Heat the wok with 2 tablespoon of oil. Add shredded pork, prawns and shiitake. Fry until fragrant. Add water and bring to a boil.
5. Add sea cucumbers, fungus, daylily and seasonings. Turn to low heat and boil for 10min. Add thickening. Keep stirring and add egg pulp. Lastly sprinkle sesame oil.
6. Bring water to a boil. Add noodles and use chopsticks to stir. Boil for 5-6min. Put noodles in 4 bowls separately. Pour in halogen gravy. Serve hot.

Longevity Noodles
長壽麵 (P100)

Time: 40 mins   Serve: 1

Ingredients

260gm Cold Water Dough (please refer to page 17), Enough Sesame oil, added lastly, 1 stalk Chinese celery, minced

Soup

20gm Lean pork, 2 Tomatoes, 4 slices Soaked fungus, 6 Fresh shrimps, 250ml Broth, 1 Egg, beaten, added lastly

Seasonings

Enough salt, Enough abalone sauce, 2 tablespoon mixture of corn starch and water.

Procedures

1. Rub and knead Cold Water Dough. Roll up to a cylinder. Roll the cylinder into a long thin. Dust with flour. Set aside.
2. Heat the wok, fan-fry shredded pork and shrimps until fragrant. Add tomatoes, keep frying until cooked.
3. Bring broth and fungus to a boil. Add seasonings. Keep stirring and pour in egg pulp. Turn the heat off.
4. Cook noodles until done. Place in the bowls.
5. Pour soup in the bowls. Sprinkle sesame oil and Chinese celery. Serve hot.

Chinese Orecchiette and Pearl Noodles
貓耳朵珍珠麵 (P102)

Time: 30 mins   Serve: 4-6

Ingredients

500gm Cold Water Dough (please refer to page 17), 50gm Plain flour, Enough sesame oil

Three treasures Soup

400gm Shredded lean pork, 8-10 pieces of Soaked fungus, 3 Soaked and prepared Sea cucumber, 30 fresh Shrimps, 6 Soak shiitakes, 1500ml Broth, 3 Eggs, beaten, 6 pieces Chinese lettuces

Seasonings

Enough Salt, Enough Abalone sauce, Enough Dark soy sauce

Thickening

2 teaspoon Corn starch, 3 tablespoon Water

Procedures

1. Divide Cold Water Dough into 2 chunks. First chunk roll into a tube about less than 1cm thick. Cut into small pieces. Place your floured thumb at the top of each piece. Use a little pressure to pull it across the piece creating kitten's ear shape.
2. Roll the second chunk of dough flat to thin square. Dust with flour. Fold 3 times. Use a knife and cut into stripes. Dust with flour again. Cut into dices to form pearl shapes.
3. Pan-fry shredded pork and shrimps. Add in broth. Bring to a boil. Add seasonings, and then fungus, sea cucumber and shiitake. Pour in thickening. Keep stirring and pour in egg pulp.
4. Bring water to a boil and cook 2 kinds of noodles separately until done.
5. Blanch lettuces in the pot of water which has cooked noodles. Remove lettuces and put in bowls of soup with noodles. Serve hot.

Mutton Soup Noodles flavoured with Sichuan Pepper Oil
羊肉熗湯麵 (P104)

Time: 20 mins   Serve: 4-6

Ingredients

1200gm Cold Water Dough (please refer to page 17)

Mutton soup

600gm Leg of lamb meat, shredded, 1000gm White Tianjin cabbages, 400gm Soaked Cellophane noodles, 200gm Chinese celery, cut into sections, 4 tablespoon Sweet bean paste, 2 bird eyes, 15gm Sichuan peppers, 1000ml Broth

Seasonings

2 teaspoon Salt, 2 tablespoon Vinegar

Procedures

1. Use rolling pin, roll Cold Water Dough flat into thin square. Cut into stripes.
2. Cook noodles. Rinse with cold tap running water.
3. Heat wok with oil. Fry Sichuan peppers until fragrant. Remove seeds.
4. Add shredded mutton and sweet bean paste in the wok. Fry until fragrant. Add Tianjin cabbages and rest of ingredients except Chinese celery. Stir fry a little. Pour in broth.
5. Add seasonings and mix well. Add cooked noodles. Sprinkle Chinese celery. Done and serve hot.

Saozi Mian
臊子麵 (P106)

Time: 40 mins   Serve: 4-6

Ingredients

Extra fine handmade noodles, 600gm Cold Water Dough (please refer to page 17), 30gm Plain flour

Soup

1 Bean curd, 400gm Shredded mutton, 6 Soaked fungus, 50gm Soaked daylily, 4 stalks Chinese celery, 6 Soaked shiitake, 20gm Paprika, 20gm Sichuan pepper, 40gm Chinese pickled cabbage, minced, 2 Eggs, beaten, 1500ml Broth

Seasonings

2 teaspoon Salt, 2 teaspoon MSG, 1/2 tablespoon Dark soy sauce, for colouring, 3 tablespoon mixture of corn starch with water, 1/2 tablespoon Sesame oil

Procedures

1. Use rolling pin, roll Cold Water Dough flat into thin square. Cut into thin stripes. Dust with flour. Set aside.
2. Cut bean curd into 1cm cubes. Rinse Chinese celery, cut into sections. Slice shiitake.
3. Bring water to a boil. Cook noodles. Keep stirring in one direction until cooked. Drain and rinse under cold running tap water. Set aside.
4. Heat wok with oil. Fry Sichuan pepper, mutton and paprika until fragrant. Pour in broth and bring to a boil. Add bean curd, shiitake, fungus and pickled cabbage. Cook until done.
5. Season with salt and MSG. Add dark soy sauce. Bring to a boil again. Add thickening. Then pour in egg pulp while stirring soup. Sprinkle sesame oil. Add Chinese celery. Place noodles inside. Stir a little until boiled. Turn off heat.

Beijing Fried Noodles
北京炒麵 (P108)

Time: 10 mins   Serve: 4-6

Ingredients

1200gm Handmade noodles

Stuffing

400gm Yellow chives, 400gm Shredded pork, 200gm Fresh shrimps, 100ml Broth

Handmade noodles

800gm flour, 350ml water, 1tsp salt

Seasonings

1.5 teaspoon Salt, 1 tablespoon Abalone sauce

Procedures

1. Handmade noodles kneading method is the same as cold water dough (please refer to page 17). Put aside and rest for 20 minutes. Roll to flat with a rolling pin. Fold to cut in strips.
2. Bring water to a boil. Cook noodles. Drain and rinse under cold running tap water.
3. Wash yellow chives. Cut into 1 inch sections.
4. Heat wok with oil. Fry yellow chives, pork and shrimps until fragrant.
5. Pour in broth. Add noodles. Stir fry. Add in seasonings. Dish up.

Eight Treasures Glutinous Rice
八寶糯米飯 (P110)

Time: 60 mins   Serve: 10

Ingredients

600gm Glutinous rice, 10 Cooked chestnuts, peeled, 3 slices Canned pineapples, 500gm Red date paste, 40gm Boiled red dates, 200gm Dried raisins, 1 Dried preserved citrus, 10 Reed leaves, 35 Pinto beans, cooked, 200gm Granulated sugar

Procedures

1. Soak glutinous rice for 6 hours. Steam for 20 min.
2. Make a round disc with 350gm in 20cm diameter. Use red date paste to make another disc, in 20cm diameter and place it on top of glutinous rice disc.
3. Use the rest of glutinous rice to make a smaller disc, to cover the red date disc. Place a slice of round pineapple on top.
4. Quarter other 2 pieces of pineapples. Insert separately in the glutinous cake.
5. In between each piece of pineapple, place a chestnut and red date.
6. Cut the citrus into desirable shape. Place on top pinto beans in each alternate red date. Cover with plastic cling. Place rice cake in a similar size bowl. Turn over and cover with reed leaves. Steam for 30min.
7. Mix sugar with water. Bring to a boil and continue to cook until thicken. Pour onto the rice cake before serving.

跟大廚做北方麵點 Northern Chinese Dumplings

著者 Author
李起發 Li Hei Fat

編輯 Editor
郭麗眉 Cecilia Kwok

翻譯 Translator
區敏華 Patricia Au

攝影 Photographer
幸浩生 Johnny Han

封面設計 Cover Designer
王妙玲 ML Wong

版面設計 Designer
辛紅梅 Cindy Xin

出版者 Publisher
萬里機構・飲食天地出版社 Food Paradise Publishing Co., an imprint of Wan Li Book Company Limited
香港鰂魚涌英皇道1065號東達中心1305室 Room 1305, Eastern Centre, 1065 King's Road, Quarry Bay, Hong Kong.
電話 Tel: 2564 7511
傳真 Fax: 2565 5539
網址 Web Site: http://www.wanlibk.com

發行者 Distributor
香港聯合書刊物流有限公司 SUP Publishing Logistics (HK) Ltd.
香港新界大埔汀麗路36號中華商務印刷大廈3字樓 3/F, C & C Building, 36 Ting Lai Road, Tai Po, N.T., Hong Kong.
電話 Tel: 2150 2100
傳真 Fax: 2407 3062
電郵 E-mail: info@suplogistics.com.hk

承印者 Printer
美雅印刷製本有限公司 Elegance Printing & Book Binding Co Ltd.

出版日期 Publishing Date
二〇一二年九月第一次印刷 First Print in September 2012
二〇一八年五月第三次印刷 Third Print in May 2018

萬里機構

萬里 Facebook

myCOOKey.com